Annals of Mathematics Studies

Number 75

THE NEUMANN PROBLEM FOR THE CAUCHY-RIEMANN COMPLEX

BY

G. B. FOLLAND AND J. J. KOHN

PRINCETON UNIVERSITY PRESS
AND
UNIVERSITY OF TOKYO PRESS

PRINCETON, NEW JERSEY
1972

LC Card: 72-1984

ISBN: 0-691-08120-4

AMS 1970: primary 35N15
secondary 58G05, 35H05, 35D10, 35F15

Published in Japan exclusively by
University of Tokyo Press
in other parts of the world by
Princeton University Press

Printed in the United States of America

FOREWORD

This book is based on the notes from lectures given by the second author at Princeton University in the year 1970-71, which were subsequently expanded and revised by the first author. Our aim has been to provide a thorough and coherent account of the solution of the $\bar{\partial}$-Neumann problem and certain of its applications and ramifications, and we have taken the opportunity afforded by the monograph format to employ a somewhat more leisurely style than is common in the original journal articles. It is our hope that this book may thereby be accessible to a fairly wide audience and that it may also provide a sort of working introduction to some of the recent techniques in partial differential equations.

In keeping with this philosophy, we have tried to make the book as self-contained as possible. On the geometrical side, we assume the reader is familiar with differentiable manifolds and their native flora and fauna: vector fields, differential forms, partitions of unity, etc. On the analytical side, we assume only an elementary knowledge of functional analysis and Fourier analysis. In the body of the text we also assume an acquaintance with Sobolev spaces and pseudodifferential operators, but we have included an appendix which develops these theories as far as they are needed.

G. B. F.
J. J. K.

JUNE, 1972

TABLE OF CONTENTS

The Neumann Problem for the
Cauchy-Riemann Complex

CHAPTER I

FORMULATION OF THE PROBLEM

1. *Introduction*

In the nineteenth century two approaches to the theory of functions of a complex variable were initiated by Weierstrass and Riemann, respectively. The first was to study power series, canonical products, and such, staying within the analytic category; the second was to work in the $\mathcal{C}^\infty$ category, using the differential equations and associated variational problems arising from the situation. The first approach was generalized to functions of several variables by K. Oka, H. Cartan, and others, and it is along these lines that the modern theory of several complex variables has largely developed. The second approach has been used with great success in the case of compact complex manifolds (the Hodge theory, cf. Weil [46]), and more recently these methods have been extended to open manifolds. This extension, however, poses rather delicate analytical problems. In particular, it leads to a non-coercive boundary value problem for the complex Laplacian, the *$\bar\partial$-Neumann problem.* It is our purpose here to present a detailed solution of this problem for domains with smooth boundary satisfying certain pseudoconvexity conditions and to indicate its applications to complex function theory.

By way of introduction, let us consider functions (and, more generally, differential forms) on a bounded domain M in $\mathbf{C}^n$ with smooth boundary bM. The Cauchy-Riemann operator $\bar\partial$ defined on functions by $\bar\partial f = \Sigma_1^n \frac{\partial f}{\partial \bar z_i} d\bar z_i$ extends naturally to yield the *Dolbeault complex*

$$0 \longrightarrow \Lambda^{p,0}(M) \xrightarrow{\bar\partial} \Lambda^{p,1}(M) \xrightarrow{\bar\partial} \dots \xrightarrow{\bar\partial} \Lambda^{p,n}(M) \longrightarrow 0$$

where $\Lambda^{p,q}(M)$ is the space of smooth forms of type (p,q) on M. The holomorphic functions are precisely the solutions of the homogeneous equation $\bar\partial f = 0$.

The inhomogeneous equation $\bar\partial f = \phi$ is also of interest. Consider the following version of the *Levi problem*: given $p \in bM$, is there a holomorphic function on M that blows up at p? In general, the answer is no: for example, if M is the region between two concentric spheres, Hartogs' theorem [18] says that any holomorphic function on M extends holomorphically to the interior of the inner sphere. However, if M is strongly convex at p (meaning that there is a neighborhood U of p such that for any $q \in (M \cup bM) \cap U$, the line segment between p and q lies in M), a classical construction of E. E. Levi guarantees the existence of a neighborhood V of p and a holomorphic function w on $M \cap V$ which blows up only at p. Now suppose the equation $\bar\partial f = \phi$ (where ϕ satisfies the compatibility condition $\bar\partial\phi = 0$) is always solvable in M in such a way that f is smooth up to bM (i.e., can be extended smoothly across bM) whenever ϕ is. Then we can solve the Levi problem. Indeed, let ψ be a smooth function with support in V which is identically one near p. Then ψw is defined on all of M and is smooth away from p; since $\bar\partial(\psi w) = 0$ near p, $\bar\partial(\psi w)$ is smooth up to the boundary. Therefore there is a function f, smooth up to the boundary, which satisfies $\bar\partial f = \bar\partial(\psi w)$. Finally, $f - \psi w$ is holomorphic in M and blows up at p. (We shall discuss this construction in greater detail in §4.2.)

Let us consider the equation $\bar\partial f = \phi$ where f and ϕ are supposed square-integrable. If a solution f exists, it is determined only modulo the space $\mathcal{H} = \{g \in L^2(M): \bar\partial g = 0\}$. By general Hilbert space theory, $\mathcal{H}^\perp$ is the closure of the range of the Hilbert space adjoint $\bar\partial^*$ of $\bar\partial$. Thus we are led to study the equation

$$\bar\partial\bar\partial^*\theta = \phi \qquad (\bar\partial\phi = 0)\,. \tag{1.1.1}$$

For general ϕ, in analogy with the deRham-Hodge construction for the exterior derivative, the proper equation is

$$(\bar\partial\bar\partial^* + \bar\partial^*\bar\partial)\,\theta = \phi \,. \tag{1.1.2}$$

(Note that (1.1.2) reduces to (1.1.1) when $\bar\partial\phi = 0$, for then

$$\bar\partial\bar\partial^*\bar\partial\theta = 0 \implies (\bar\partial\bar\partial^*\bar\partial\theta, \bar\partial\theta) = 0 \implies (\bar\partial^*\bar\partial\theta, \bar\partial^*\bar\partial\theta) = 0 \implies \bar\partial^*\bar\partial\theta = 0 \text{ .)}$$

The equation (1.1.2) is a boundary value problem in disguise, for $\bar\partial^*$ is a differential operator obtained from the equation $(\bar\partial^*\phi, \psi) = (\phi, \bar\partial\psi)$ by formal integration by parts, and the forms in the domain of $\bar\partial^*$ must therefore satisfy conditions which guarantee that the boundary terms in the integration by parts always vanish. (1.1.2) may also be considered as a variational problem, cf. Morrey [34].

Thus, with much hand-waving and little precision, we have set up the *$\bar\partial$-Neumann problem*: prove existence and regularity for solutions of (1.1.2). The $\bar\partial$-Neumann problem was proposed by D. C. Spencer in the early 1950's as a means of extending Hodge theory to open manifolds and obtaining existence theorems for holomorphic functions; he also pioneered the generalization of this approach in the study of more general overdetermined systems. Related boundary value problems were studied by Garabedian and Spencer [12], Kohn and Spencer [29], and Conner [8] using integral operators, but these methods were not powerful enough to solve the $\bar\partial$-Neumann problem. Later Morrey [33] introduced the "basic estimate," and the problem was solved by Kohn [22] by establishing regularity. However, the regularity proof in [22] has since been supplanted by a better proof using the technique of elliptic regularization developed by Kohn and Nirenberg [27]. It is the latter method which we shall employ here; another version of this proof may be found in the book of Morrey [34]. A different approach has been developed by Hormander [16], [18]; we shall discuss his work briefly in §6.1.

We shall now retrace our steps with more care and in greater generality. The natural setting for the $\bar\partial$-Neumann problem is the class of compact complex manifolds with boundary. However, we shall go one step further and work with integrable almost-complex manifolds so that we will be in a position to prove the Newlander-Nirenberg theorem to the effect that every integrable almost-complex manifold is in fact complex. This presents no additional complications, and the reader who wishes to envision all our manifolds as complex is free to do so.

2. *Almost-complex manifolds and differential operators*

Let M be a real $\mathcal{C}^\infty$ manifold of dimension m. An *almost-complex structure* on M is a splitting of the complexified tangent bundle $CTM = TM \otimes_R C$ by projections $\Pi_{1,0}$ and $\Pi_{0,1}$ such that $\Pi_{1,0} + \Pi_{0,1} = 1$, $\Pi_{1,0}\Pi_{0,1} = \Pi_{0,1}\Pi_{1,0} = 0$, and $\Pi_{0,1} = \overline{\Pi_{1,0}}$. (The last equation means that for $\xi \in CTM$, $\Pi_{0,1}\xi = \overline{(\Pi_{1,0}\bar{\xi})}$ where $\overline{}$ denotes complex conjugation.) We write $T_{1,0}M = \text{Range}(\Pi_{1,0})$ and $T_{0,1}M = \text{Range}(\Pi_{0,1})$; note that $\dim_C T_{1,0}M = \dim_C T_{0,1}M = \frac{m}{2}$, so m must be even; we write $m = 2n$. One can easily verify that an almost-complex structure induces a preferred orientation on M, by restricting the coordinate transformations to those which preserve $\Pi_{1,0}$ and $\Pi_{0,1}$.

The projections $\Pi_{1,0}$ and $\Pi_{0,1}$ naturally induce a splitting of the exterior powers of the complexified cotangent bundle, $\Lambda^k CT^*M = \bigoplus_{\substack{p+q=k \\ 0 \le p,q \le n}} \Lambda^{p,q} CT^*M$, and we denote the projection $\Lambda^k CT^*M \to \Lambda^{p,q} CT^*M$ by $\Pi_{p,q}$. The space of $\mathcal{C}^\infty$ sections of $\Lambda^{p,q}CT^*M$, i.e., the *forms of type* (p,q) on M, will be denoted by $\Lambda^{p,q}(M)$. We define the operators $\partial: \Lambda^{p,q}(M) \to \Lambda^{p+1,q}(M)$ and $\bar{\partial}: \Lambda^{p,q}(M) \to \Lambda^{p,q+1}(M)$ by $\partial\phi = \Pi_{p+1,q} d\phi$, $\bar{\partial}\phi = \Pi_{p,q+1} d\phi$. Since $d = \bar{d}$ and $\overline{\Pi_{p,q}} = \Pi_{q,p}$, we have $\bar{\partial}\phi = \overline{(\partial\bar{\phi})}$. It is clear from the corresponding properties of d that ∂ and $\bar{\partial}$ act locally and satisfy the derivation law:

$$\bar{\partial}(\phi \wedge \psi) = \bar{\partial}\phi \wedge \psi + (-1)^{p+q}\phi \wedge \bar{\partial}\psi \quad \text{for} \quad \phi \in \Lambda^{p,q}(M) .$$

The *torsion tensor* of the almost-complex structure is the bilinear map T from complex vector fields to complex vector fields defined by

$$T(X, Y) = \Pi_{1,0}[\Pi_{0,1}X, \Pi_{0,1}Y] + \Pi_{0,1}[\Pi_{1,0}X, \Pi_{1,0}Y] .$$

(1.2.1) PROPOSITION. *The following properties are equivalent*:

(1) (1) $T = 0$; (2) $\bar{\partial}^2 = 0$; (3) $\partial^2 = 0$; (4) $d = \partial + \bar{\partial}$.

Proof: That (2) $\Longleftrightarrow$ (3) is obvious. That (4) $\Longrightarrow$ (2) and (3) follows immediately from the equation $d^2 = 0$ and the fact that forms of different type are linearly independent. To show that (1) $\Longleftrightarrow$ (4), we use the identity

$$2d\phi(X, Y) = X\phi(Y) - Y\phi(X) - \phi([X, Y])$$

for one-forms ϕ. If X, Y are sections of $T_{1,0}M$, (1) implies that $[X, Y]$ is also, so $d\phi(X, Y) = 0$ for $\phi \in \Lambda^{0,1}(M)$. Thus $d(\Lambda^{0,1}(M)) \subset \Lambda^{1,1}(M) + \Lambda^{0,2}(M)$, so $d = \partial + \bar{\partial}$ on $\Lambda^{0,1}(M)$. Similarly $d = \partial + \bar{\partial}$ on $\Lambda^{1,0}(M)$. But $d = \partial + \bar{\partial}$ trivially on functions, so by the derivation law and the fact that all forms are locally products of functions and one-forms, $d = \partial + \bar{\partial}$ everywhere. The implication (4) $\Longrightarrow$ (1) follows by reversing this argument. Finally, if (2) holds, from the definition of $\bar{\partial}$ we have for any sections X, Y of $T_{0,1}M$ and any function f,

$$\begin{aligned} 0 &= 2\bar{\partial}^2 f(X, Y) = 2d\bar{\partial}f(X, Y) \\ &= X\bar{\partial}f(Y) - Y\bar{\partial}f(X) - \bar{\partial}f([X, Y]) \\ &= XYf - YXf - (\Pi_{0,1}[X, Y])f\,, \end{aligned}$$

and hence $[X, Y] = \Pi_{0,1}[X, Y]$. Likewise, by (3), $\Pi_{1,0}[X, Y] = [X, Y]$ for all sections X, Y of $T_{1,0}M$. Thus (1) holds. Q.E.D.

An almost-complex structure satisfying the conditions of Proposition (1.2.1) is called *integrable*. Condition (1) says that the sections of $T_{1,0}M$ and $T_{0,1}M$ form Lie algebras, i.e., they are integrable distributions in the sense of Frobenius. Condition (2) says that the sequence

$$(1.2.2) \qquad 0 \longrightarrow \Lambda^{p,0}(M) \xrightarrow{\bar{\partial}} \Lambda^{p,1}(M) \xrightarrow{\bar{\partial}} \cdots \xrightarrow{\bar{\partial}} \Lambda^{p,n}(M) \longrightarrow 0$$

is a *complex*. This is the property which will be crucial for our purposes.

If M is actually a complex manifold, that is, it possesses a covering by charts with complex coordinates $\{z_j = x_j + iy_j\}_1^n$ which are holomorphically related, then M has a natural almost-complex structure. In local coordinates, if we set $\frac{\partial}{\partial z_j} = \frac{1}{2}\left(\frac{\partial}{\partial x_j} - i\frac{\partial}{\partial y_j}\right)$, $\frac{\partial}{\partial \bar{z}_j} = \frac{1}{2}\left(\frac{\partial}{\partial x_j} + i\frac{\partial}{\partial y_j}\right)$, the structure is given by

$$\Pi_{1,0}\left(\sum\left(a_j\frac{\partial}{\partial z_j} + b_j\frac{\partial}{\partial \bar{z}_j}\right)\right) = \sum a_j\frac{\partial}{\partial z_j},$$

$$\Pi_{0,1}\left(\sum\left(a_j\frac{\partial}{\partial z_j} + b_j\frac{\partial}{\partial \bar{z}_j}\right)\right) = \sum b_j\frac{\partial}{\partial \bar{z}_j}.$$

That this is independent of the coordinate representation follows from the Cauchy-Riemann equations. It is easily verified that this structure is integrable.

A *Hermitian metric* on an almost-complex manifold M is a Hermitian inner product $< , >_x$ on each $\Pi_{1,0}(CT_xM)$ varying smoothly in x. For $\xi, \eta \in CT_xM$, we set

$$<\xi, \eta>_x = <\Pi_{1,0}\xi, \Pi_{1,0}\eta>_x + \overline{<\Pi_{1,0}\bar{\xi}, \Pi_{1,0}\bar{\eta}>_x}.$$

The inner product $< , >_x$ then extends naturally to all the spaces $\Lambda^{p,q}CT_x^*M$. If $\omega_1, \ldots, \omega_n$ is an orthonormal basis for $\Lambda^{1,0}CT_x^*M$, then $\omega_1 \wedge \bar{\omega}_1 \wedge \omega_2 \wedge \bar{\omega}_2 \wedge \ldots \wedge \omega_n \wedge \bar{\omega}_n = \gamma$ is the *volume element* on M at x. We define global scalar products for forms by $(\phi, \psi) = \int_M <\phi, \psi>\gamma$ for $\phi, \psi \in \Lambda^{p,q}(M)$. (We shall generally omit writing γ in such integrals.)

If M is a Hermitian almost-complex manifold, the *formal adjoint* ϑ of $\bar{\partial}$ is the differential operator from $\Lambda^{p,q}(M)$ to $\Lambda^{p,q-1}(M)$ defined by $(\vartheta\phi, \psi) = (\phi, \bar{\partial}\psi)$ for all $\psi \in \Lambda^{p,q-1}(M)$ with compact support. If M is integrable, clearly $\vartheta^2 = 0$. The operator $\Box = \bar{\partial}\vartheta + \vartheta\bar{\partial}$ is called the *complex Laplacian*. If M is a Kähler manifold, which means that M is complex and the (1, 1)-form Ω defined by $\Omega(X, Y) = <X, i(\Pi_{1,0} - \Pi_{0,1})Y>$ is closed, one can show (cf. [46]) that $\Box = \frac{1}{2}\Delta$ where Δ is the ordinary Laplacian of the deRham complex (= the negative of the classical Laplacian on functions).

Let us work out the coordinate formulas for $\bar\partial$ and ϑ for the case $M = C^n$ (with the Euclidean metric). Every form $\phi \in \Lambda^{p,q}(C^n)$ is expressed as $\phi = \Sigma_{IJ}\, \phi_{IJ}\, dz^I \wedge d\bar z^J$ where I and J are strictly increasing sequences of positive integers of length p (resp. q). (Thus, for example, if $I = (i_1, \ldots, i_p)$, $dz^I = dz_{i_1} \wedge \ldots \wedge dz_{i_p}$). Then

$$\bar\partial\phi = \sum_{kIJ} \frac{\partial\phi_{IJ}}{\partial \bar z_k} d\bar z_k \wedge dz^I \wedge d\bar z^J$$

$$= (-1)^p \sum_{kIJK} \epsilon^K_{kJ} \frac{\partial\phi_{IJ}}{\partial \bar z_k} dz^I \wedge d\bar z^K$$

where ϵ^K_{kJ} is the sign of the permutation changing $(k, j_1, \ldots, j_q)$ into $K = (k_1, \ldots, k_{q+1})$ (where $k_1 < \ldots < k_{q+1}$) if $\{k, j_1, \ldots, j_q\} = \{k_1, \ldots, k_{q+1}\}$ as sets, and is zero otherwise.

To compute ϑ we use the fact that $\int_{C^n} u \overline{\frac{\partial v}{\partial \bar z_k}} = - \int_{C^n} \frac{\partial u}{\partial z_k} \bar v$ when v has compact support. (Proof: integration by parts.) Thus if ϕ is as above, and $\psi = \Sigma_{IH} \psi_{IH}\, dz^I \wedge d\bar z^H \in \Lambda^{p,q-1}(M)$ has compact support,

$$(\phi, \bar\partial\psi) = 2^{p+q}(-1)^p \sum_{IJ} \left(\phi_{IJ}, \sum_{kH} \epsilon^J_{kH} \frac{\partial\psi_{IH}}{\partial \bar z_k}\right)$$

$$= 2^{p+q}(-1)^{p+1} \sum_{IH} \left(\sum \epsilon^J_{kH} \frac{\partial\phi_{IJ}}{\partial z_k}, \psi_{IH}\right).$$

(The annoying factor of 2^{p+q} comes from the fact that $\langle dz_i, dz_j\rangle = 2\delta_{ij}$.) Thus

$$\vartheta\phi = 2(-1)^{p+1} \sum_{IHJk} \epsilon^J_{kH} \frac{\partial\phi_{IJ}}{\partial z_k} dz^I \wedge d\bar z^H .$$

On a general manifold M these formulas remain valid if we replace $dz_1, \ldots, dz_n$ by a local orthonormal basis $\omega_1, \ldots, \omega_n$ for $\Lambda^{0,1}(M)$, except for some error terms resulting from the fact that the forms ω_j are not $\bar\partial$-closed and the coefficients of the metric are not constant. These terms, however, do not involve differentiation of the coefficients of the forms. Thus we have, for $\phi \in \Lambda^{p,q}(M)$,

(1.2.3) $\bar{\partial}\phi = (-1)^p \sum_{kIJK} \epsilon^K_{kJ} \bar{L}_k(\phi_{IJ})\omega^I \wedge \bar{\omega}^K$ + terms of order zero,

(1.2.4) $\vartheta\phi = (-1)^{p+1} \sum_{kIHJ} \epsilon^J_{kH} L_k(\phi_{IJ})\omega^I \wedge \bar{\omega}^H$ + terms of order zero,

where L_k is the vector field dual to ω_k. (There is no factor of 2 here because the ω_k's are orthonormal!)

In general, if E and F are vector bundles on a manifold M and $D: \Gamma(E) \to \Gamma(F)$ is a differential operator of order k, where Γ denotes spaces of sections, we define for each $\eta \neq 0$ in T^*_xM the *symbol* $\sigma(D, \eta): E_x \to F_x$ of D at η as follows. Let ρ be a function defined near x with $\rho(x) = 0$ and $d\rho_x = \eta$, and for each $\theta \in E_x$ let $\tilde{\theta}$ be a local section of E extending θ. Then $\sigma(D, \eta)\theta = \frac{1}{k!} D(\rho^k\tilde{\theta})|_x$. It is easily verified that $\sigma(D, \eta)\theta$ is independent of the choice of ρ and $\tilde{\theta}$. Moreover, if D' is the formal adjoint of D with respect to Hermitian structures on E and F, $\sigma(D', \eta) = (-1)^k \sigma(D, \eta)^*$; and if $D_1: \Gamma(F) \to \Gamma(G)$ is another differential operator, $\sigma(D_1 D, \eta) = \sigma(D_1, \eta)\, \sigma(D, \eta)$. If $D_1 D = 0$ and the symbol sequence $E_x \xrightarrow{\sigma(D,\eta)} F_x \xrightarrow{\sigma(D,\eta)} G_x$ is exact for all $x \in M$ and $\eta \neq 0 \in T^*_xM$, the sequence $\Gamma(E) \xrightarrow{D} \Gamma(F) \xrightarrow{D_1} \Gamma(G)$ is called an *elliptic complex*. A single operator D is called *elliptic* if $0 \longrightarrow \Gamma(E) \xrightarrow{D} \Gamma(F)$ is an elliptic complex, i.e., if its symbol is injective. It is an easy exercise to prove that $\Gamma(E) \xrightarrow{D} \Gamma(F) \xrightarrow{D_1} \Gamma(G)$ is an elliptic complex if and only if $D'D + D_1 D_1' : \Gamma(F) \to \Gamma(F)$ is elliptic. For details on these matters, see, e.g., Palais [38].

Let us compute the symbols of $\bar{\partial}$ and ϑ on an integrable Hermitian almost-complex manifold M. With η, ρ as above and $\phi \in \Lambda^{p,q}(M)$, $\sigma(\bar{\partial}, \eta)\phi_x = \bar{\partial}(\rho\phi)_x = [\bar{\partial}\rho \wedge \phi + \rho\bar{\partial}\phi]_x = \Pi_{0,1}\eta \wedge \phi_x$, so $\sigma(\bar{\partial}, \eta) = \Pi_{0,1}\eta \wedge (\cdot)$. Therefore, $\sigma(\vartheta, \eta) = -\,\Pi_{0,1}\eta \vee (\cdot)$ where $\langle\alpha \wedge \beta, \gamma\rangle = \langle\beta, \alpha \vee \gamma\rangle$. (The reader who wishes to have the formula for $\sigma(\vartheta, \eta)$ involve $\Pi_{1,0}\eta$ instead of $\Pi_{0,1}\eta$ must adjust the definition of $\vee$ to make peace with the conjugate linearity lurking in these duality relations.)

Since η is real, $\Pi_{0,1}\eta \neq 0$ whenever $\eta \neq 0$. Thus if θ is a (p, q)-covector at x satisfying $(\Pi_{0,1}\eta) \wedge \theta = 0$, by expressing θ in terms of a basis for CT_x^*M of which $\Pi_{0,1}\eta$ is a member, we see that θ is divisible by $\Pi_{0,1}\eta$, i.e., $\theta = \Pi_{0,1}\eta \wedge \psi$ for some (p, q−1)-covector ψ. Therefore:

(1.2.5) PROPOSITION. *The symbol sequence for* $\bar\partial$ *is exact, i.e., the* $\bar\partial$ *complex* (1.2.2) *is elliptic.*

One could also show this by verifying that $\sigma(\Box, \eta)\theta = \sigma(\bar\partial\vartheta + \vartheta\bar\partial, \eta)\theta = -|\Pi_{0,1}\eta|^2\theta$, so $\Box$ is elliptic.

3. *Operators on Hilbert space*

From now on we shall be working on a Hermitian integrable almost-complex manifold M of real dimension 2n with smooth boundary bM such that $\overline{M} = M \cup bM$ is compact. We shall assume, without loss of generality, that M is imbedded in a slightly larger open manifold M′ and that bM is defined by the equation $r = 0$ where r is a real $\mathcal{C}^\infty$ function with $r < 0$ inside M, $r > 0$ outside $\overline{M}$, and $|dr| = 1$ on bM. (r may be constructed from the geodesic distance to bM.) We introduce some spaces of forms and distributions.

$\Lambda^{p,q}(M)$ is, as before, the space of $\mathcal{C}^\infty$ (p, q)-forms on M. $\Lambda^{p,q}(\overline{M})$ is the subspace of $\Lambda^{p,q}(M)$ whose elements can be extended smoothly to M′. $\Lambda_0^{p,q}(M)$ is the subspace of $\Lambda^{p,q}(\overline{M})$ whose elements have compact support disjoint from bM. For $s \in Z$, $H_s^{p,q}$ is the Sobolev space of distribution-valued (p, q)-forms on $\overline{M}$ of order s, and $\| \ \|_s$ denotes any of the equivalent norms defining this space. (For precise definitions and the basic properties of $H_s^{p,q}$, see the Appendix, §1-2.) In particular, $H_0^{p,q}$ is the space of square-integrable (p, q)-forms on M, and we shall generally denote the norm in $H_0^{p,q}$ simply by $\| \ \|$. Also, if s is a positive integer, $H_s^{p,q}$ is the space of (p, q)-forms whose coefficients have weak L^2 derivatives up to order s; if $\phi \in H_s^{p,q}$ has support in a coordinate patch with coordinates $x_1, \ldots, x_{2n}$, say $\phi = \Sigma_I \phi_I dx^I$, we may take $\|\phi\|_s^2 = \Sigma_{0 \le |\alpha| \le s} \|D^\alpha \phi_I\|^2$ where $D^\alpha = \left(\frac{1}{i}\frac{\partial}{\partial x_1}\right)^{\alpha_1} \cdots \left(\frac{1}{i}\frac{\partial}{\partial x_{2n}}\right)^{\alpha_{2n}}$, $|\alpha| = \Sigma_1^{2n} \alpha_i$.

We shall henceforth use the symbol $\bar\partial$ to mean the closure of $\bar\partial|\Lambda^{p,q}(\bar M)$ with respect to $H_0^{p,q}$, in other words, the operator whose graph is the closure of the graph of $\bar\partial|\Lambda^{p,q}(\bar M)$ in $H_0^{p,q} \times H_0^{p,q+1}$ (cf. Riesz-Sz. Nagy [39]). Our next task will be to determine the relation between the Hilbert space adjoint $\bar\partial^*$ of $\bar\partial$ and its formal adjoint ϑ.

Suppose L is a vector field on M', and let L' be its formal adjoint on M defined by $(Lu, v) = (u, L'v)$ for all $v \in \Lambda_0^{0,0}(M)$. We shall show how this equation must be modified when v does not have compact support. By using a partition of unity we may assume that u and v are supported in a coordinate patch in M'. If the patch lies inside M, then of course $(Lu, v) = (u, L'v)$. If it intersects bM, we choose coordinates $t_1, \ldots, t_{2n-1}, r$ where r is the function defining bM. Let $\gamma = G(t_1, \ldots, t_{2n-1}, r)\, dt_1 \wedge \ldots dt_{2n-1} \wedge dr$ be the volume form. Then we may write $L = \Sigma_1^{2n-1} a_j \frac{\partial}{\partial t_j} + b\frac{\partial}{\partial r}$, and

$$
\begin{aligned}
(Lu,v) &= \sum \int_M a_j \frac{\partial u}{\partial t_j}\bar v\gamma + \int_M b\frac{\partial u}{\partial r}\bar v\gamma \\
&= \sum \int_{-\infty}^{0}\int_{R^{2n-1}} a_j \frac{\partial u}{\partial t_j}\bar v G\, dt\, dr + \int_{-\infty}^{0}\int_{R^{2n-1}} b\frac{\partial u}{\partial r}\bar v G\, dt\, dr \\
&= -\sum \int_{-\infty}^{0}\int_{R^{2n-1}} u\frac{\partial}{\partial t_j}(a_j \bar v G)\, dt\, dr + \sum \int_{-\infty}^{0}\int_{R^{2n-1}} \frac{\partial}{\partial t_j}(a_j u\bar v G)\, dt\, dr \\
&\quad - \int_{-\infty}^{0}\int_{R^{2n-1}} u\frac{\partial}{\partial r}(b\bar v G)\, dt\, dr + \int_{-\infty}^{0}\int_{R^{2n-1}} \frac{\partial}{\partial r}(bu\bar v G)\, dt\, dr \\
&= (u,L'v) + \sum \int_{-\infty}^{0}\int_{R^{2n-1}} \frac{\partial}{\partial t_j}(a_j u\bar v G)\, dt\, dr + \int_{-\infty}^{0}\int_{R^{2n-1}} \frac{\partial}{\partial r}(bu\bar v G)\, dt\, dr \\
&= (u,L'v) + \int_{R^{2n-1}} [bu\bar v G]_{r=0}\, dt
\end{aligned}
$$

by the Fundamental Theorem of Calculus. To state this invariantly, we note that $b = \sigma(L, dr)$, so that

$$(Lu, v) = (u, L'v) + \int_{bM} \langle \sigma(L, dr)u, v \rangle .$$

Finally, this formula remains valid if L is replaced by a first-order operator from one vector bundle to another, since locally we can trivialize the bundles and express L as a matrix of scalar operators and its symbol as a corresponding matrix of scalar symbols. Therefore, in particular, we have:

(1.3.1) PROPOSITION. *For all* $\phi \in \Lambda^{p,q}(\overline{M})$, $\theta \in \Lambda^{p,q+1}(\overline{M})$, $\psi \in \Lambda^{p,q-1}(\overline{M})$,

$$(\bar\partial\phi, \theta) = (\phi, \vartheta\theta) + \int_{bM} \langle \sigma(\bar\partial, dr)\phi, \theta \rangle ,$$

$$(\vartheta\phi, \psi) = (\phi, \bar\partial\psi) + \int_{bM} \langle \sigma(\vartheta, dr)\phi, \psi \rangle .$$

Recall now that the Hilbert space adjoint $\bar\partial^*$ of $\bar\partial$ is defined on the domain $\mathrm{Dom}\,(\bar\partial^*)$ consisting of all $\phi \in H_0^{p,q}$ such that for some constant $c > 0$, $|(\phi, \bar\partial\psi)| \le c\,\|\psi\|$ for all $\psi \in \Lambda^{p,q-1}(\overline{M})$. For such a ϕ, $\psi \mapsto (\phi, \bar\partial\psi)$ extends to a bounded functional on $H_0^{p,q}$, and $\bar\partial^*\phi$ is its dual vector. We define $\mathfrak{D}^{p,q}$ to be $\mathrm{Dom}\,(\bar\partial^*) \cap \Lambda^{p,q}(\overline{M})$.

(1.3.2) PROPOSITION. $\mathfrak{D}^{p,q} = \{\phi \in \Lambda^{p,q}(\overline{M}) \colon \sigma(\vartheta, dr)\phi = 0$ *on* $bM\}$, *and* $\bar\partial^* = \vartheta$ *on* $\mathfrak{D}^{p,q}$.

Proof: If $\phi \in \mathfrak{D}^{p,q}$ and $\psi \in \Lambda^{p,q-1}(\overline{M})$,

$$(\bar\partial^*\phi, \psi) = (\phi, \bar\partial\psi) = (\vartheta\phi, \psi) - \int_{bM} \langle \sigma(\vartheta, dr)\phi, \psi \rangle .$$

If $\psi \in \Lambda_0^{p,q}(M)$ the boundary term vanishes, so $(\bar\partial^*\phi, \psi) = (\vartheta\phi, \psi)$. But

$\Lambda_0^{p,q}(M)$ is dense in $H_0^{p,q}$, so we must have $\bar\partial^*\phi = \vartheta\phi$. This forces the boundary term to be zero for all $\psi \in \Lambda^{p,q-1}(\bar{M})$, so $\sigma(\vartheta, dr)\phi = 0$ on bM. Q.E.D.

We now wish to describe a self-adjoint form of the operator $\Box$ which suits its role in the $\bar\partial$ complex. To do this we use the following version of the famous Friedrichs Extension Theorem.

Suppose H is a Hilbert space and Q is a Hermitian form defined on a dense subspace D of H satisfying $Q(\phi,\phi) \geq \|\phi\|^2$ for $\phi \in D$. Suppose further that D is a Hilbert space under the inner product Q. Then there is a canonical self-adjoint operator F on H associated with Q as follows. For each $a \in H$, $\psi \mapsto (a,\psi)$ is a Q-bounded functional on D since $(a,\psi) \leq \|a\|\,\|\psi\| \leq \|a\|\sqrt{Q(\psi,\psi)}$. Thus there is a unique $\phi \in D$ such that $Q(\phi,\psi) = (a,\psi)$ for all $\psi \in D$. Define $T: H \to D \subset H$ by $Ta = \phi$. Then $\|Ta\|^2 \leq Q(Ta, Ta) = (a, Ta) \leq \|a\|\,\|Ta\|$, so T is bounded. Also, $Ta = 0$ implies $Q(Ta,\psi) = (a,\psi) = 0$ for all $\psi \in D$; thus $a = 0$ since D is dense in H, and T is injective. Next, $(Ta,\beta) = \overline{(\beta, Ta)} = \overline{Q(T\beta, Ta)} = Q(Ta, T\beta) = (a, T\beta)$, so T is self-adjoint. Finally, we set $F = T^{-1}$ and obtain:

(1.3.3) PROPOSITION (Friedrichs). F *is the unique self-adjoint operator with* Dom (F) $\subset$ D *satisfying* $Q(\phi,\psi) = (F\phi,\psi)$ *for all* $\phi \in$ Dom (F) *and* $\psi \in D$.

(1.3.4) COROLLARY. *If the* Q*-unit ball of* D *is compact in* H, *then* $T = F^{-1}$ *is a compact operator.*

In our case, we define the form Q on $\mathfrak{D}^{p,q}$ by

$$Q(\phi,\psi) = (\bar\partial\phi, \bar\partial\psi) + (\vartheta\phi, \vartheta\psi) + (\phi,\psi)$$

and let $\tilde{\mathfrak{D}}^{p,q}$ be the completion of $\mathfrak{D}^{p,q}$ under Q. The inclusion $\mathfrak{D}^{p,q} \to H_0^{p,q}$ extends uniquely to a norm-decreasing map $\tilde{\mathfrak{D}}^{p,q} \to H_0^{p,q}$; we must show this map is injective so that we can identify $\tilde{\mathfrak{D}}^{p,q}$ with a subspace of $H_0^{p,q}$ and apply the Friedrichs construction. But if $\{\phi_n\}$ is a Q-Cauchy sequence in $\mathfrak{D}^{p,q}$, then $\{\phi_n\}$, $\{\bar\partial\phi_n\}$, and $\{\vartheta\phi_n\}$ are all Cauchy sequences in $H_0^{p,q}$. Let $\phi = \lim \phi_n$ in $H_0^{p,q}$. Since $\bar\partial$ and $\bar\partial^*$ are closed operators, we have (by Proposition (1.3.2)) $\phi \in \mathrm{Dom}(\bar\partial) \cap \mathrm{Dom}(\bar\partial^*)$ and $Q(\phi,\phi) = \|\bar\partial\phi\|^2 + \|\bar\partial^*\phi\|^2 + \|\phi\|^2$. Thus if $\phi = 0$, $\lim Q(\phi_n,\phi_n) = Q(\phi,\phi) = 0$, so $\phi_n \to 0$ in $\tilde{\mathfrak{D}}^{p,q}$.

We denote the Friedrichs operator associated to Q by F. Since for $\phi,\psi \in \Lambda_0^{p,q}(M)$, $Q(\phi,\psi) = ((\Box + I)\phi,\psi)$, we see that F is a self-adjoint extension of the Hermitian operator $(\Box+I) \mid \Lambda_0^{p,q}(M)$. The smooth elements of $\tilde{\mathfrak{D}}^{p,q}$ are described by the boundary condition $\sigma(\vartheta, dr)\phi = 0$ on bM; the smooth elements of Dom (F) are characterized by a further first-order boundary condition (the so-called "*free boundary condition*").

(1.3.5) PROPOSITION. *If* $\phi \in \mathfrak{D}^{p,q}$, *then* $\phi \in \mathrm{Dom}(F)$ *if and only if* $\bar\partial\phi \in \mathfrak{D}^{p,q+1}$, *in which case* $F\phi = (\Box+I)\phi$.

Proof: Assume $\phi \in \mathfrak{D}^{p,q} \cap \mathrm{Dom}(F)$. Then for all $\psi \in \Lambda_0^{p,q}(M)$,

$$(F\phi,\psi) = (\bar\partial\phi,\bar\partial\psi) + (\vartheta\phi,\vartheta\psi) + (\phi,\psi) = ((\Box + I)\phi,\psi) . \tag{1.3.6}$$

Since $\Lambda_0^{p,q}(M)$ is dense in $H_0^{p,q}$ we must have $F\phi = (\Box + I)\phi$, and the equation (1.3.6) must hold for all $\psi \in \mathfrak{D}^{p,q}$. Now by Propositions (1.3.1) and (1.3.2),

$$\begin{aligned}(\vartheta\phi,\vartheta\psi) &= (\bar\partial\vartheta\phi,\psi) - \int_{bM} \langle\sigma(\bar\partial, dr)\vartheta\phi,\psi\rangle \\ &= (\bar\partial\vartheta\phi,\psi) + \int_{bM} \langle\vartheta\phi,\sigma(\vartheta, dr)\psi\rangle \\ &= (\bar\partial\vartheta\phi,\psi) ,\end{aligned}$$

so this term causes no trouble. However,

$$(\bar\partial\phi,\bar\partial\psi) = (\vartheta\bar\partial\phi,\psi) - \int_{bM} \langle\sigma(\vartheta, dr)\bar\partial\phi,\psi\rangle .$$

In particular, we may take $\psi = \sigma(\vartheta, dr)\bar\partial\phi$, which is in $\mathcal{D}^{p,q}$ since $\sigma(\vartheta, dr)^2 = \sigma(\vartheta^2, dr) = 0$. Thus the boundary term will vanish only when $\sigma(\vartheta, dr)\bar\partial\phi = 0$ on bM, i.e., $\bar\partial\phi \in \mathcal{D}^{p,q+1}$. Conversely, if $\bar\partial\phi \in \mathcal{D}^{p,q+1}$ then $Q(\phi,\psi) = ((\Box + I)\phi, \psi)$ for all $\psi \in \mathcal{D}^{p,q}$; hence $\phi \in \mathrm{Dom}\,(F)$ and $F\phi = (\Box + I)\phi$. Q.E.D.

The boundary conditions $\sigma(\vartheta, dr)\phi = 0$ and $\sigma(\vartheta, dr)\bar\partial\phi = 0$ on bM are the *$\bar\partial$-Neumann conditions,* and the domain of F, or of the corresponding self-adjoint extension $F - I$ of $\Box$, consists of those forms which satisfy the $\bar\partial$-Neumann conditions in a suitable weak sense. They are precisely the right conditions to guarantee that the formal adjoint of $\bar\partial$ is its Hilbert space adjoint. There is another neat characterization of the $\bar\partial$-Neumann conditions for functions on Kähler manifolds.

(1.3.7) PROPOSITION. *If* M *is a Kähler manifold, so that* $\Box = \frac{1}{2}\Delta$, *then a function is holomorphic on* $\bar M$ *if and only if it is harmonic and satisfies the $\bar\partial$-Neumann conditions.*

Proof: The condition $\sigma(\vartheta, dr)f = 0$ is vacuous for functions. If f is holomorphic, then trivially $\sigma(\vartheta, dr)\bar\partial f = 0$ and $\Delta f = 2\vartheta\bar\partial f = 0$. Conversely if f is harmonic and satisfies $\sigma(\vartheta, dr)\bar\partial f = 0$ on bM, then $0 = (\Delta f, f) = 2(\vartheta\bar\partial f, f) = 2(\bar\partial f, \bar\partial f)$, so $\bar\partial f = 0$. Q.E.D.

In general, if $\Gamma(E) \xrightarrow{D} \Gamma(F) \xrightarrow{D} \Gamma(G)$ is an elliptic complex on a manifold with boundary M, we say that $\phi \in \Gamma(F)$ satisfies the *abstract Neumann conditions* if $(D'\phi, \psi) = (\phi, D\psi)$ for all $\psi \in \Gamma(E)$ and $(D'D\phi, \psi) = (D\phi, D\psi)$ for all $\psi \in \Gamma(F)$ where D' is the formal adjoint of D. In particular if D is exterior differentiation on forms, we obtain the coercive d-Neumann problem which was solved by Conner [8]. The study of the boundary value problem for $DD' + D'D$ defined by the Neumann conditions plays an important role in the recent and continuing development of the theory of overdetermined systems of partial differential equations, cf. Spencer [41], Sweeney [43], Kohn [24].

We remark at this point that there is another way of arriving at the operator F, which was used in [22]. The following proposition is essentially due to Gaffney [11].

(1.3.8) PROPOSITION. *Let* $F_1 = \bar\partial\bar\partial^* + \bar\partial^*\bar\partial + I$ *defined on* $\mathrm{Dom}\,(F_1) = \{\phi \in \mathrm{Dom}\,(\bar\partial) \cap \mathrm{Dom}\,(\bar\partial^*)\colon \bar\partial\phi \in \mathrm{Dom}\,(\bar\partial^*)$ *and* $\bar\partial^*\phi \in \mathrm{Dom}\,(\bar\partial)\}$. *Then* F_1^{-1} *exists, is bounded and everywhere defined; and* F_1 *is self-adjoint.*

Proof: By a theorem of von Neumann [39], $(I + \bar\partial\bar\partial^*)^{-1}$ and $(I + \bar\partial^*\bar\partial)^{-1}$ are bounded self-adjoint operators. Thus $S = (I + \bar\partial\bar\partial^*)^{-1} + (I + \bar\partial^*\bar\partial)^{-1} - I$ is bounded and self-adjoint; we shall prove the proposition by showing that $S = F_1^{-1}$. First,

$$(1.3.9) \qquad (I+\bar\partial\bar\partial^*)^{-1} - I = (I-(I+\bar\partial\bar\partial^*))(I+\bar\partial\bar\partial^*)^{-1} = -\bar\partial\bar\partial^*(I+\bar\partial\bar\partial^*)^{-1},$$

$$(1.3.10) \qquad (I+\bar\partial^*\bar\partial)^{-1} - I = (I-(I+\bar\partial^*\bar\partial))(I+\bar\partial^*\bar\partial)^{-1} = -\bar\partial^*\bar\partial(I+\bar\partial^*\bar\partial)^{-1}.$$

These equations show that $\mathrm{Range}\,(I+\bar\partial\bar\partial^*)^{-1} \subset \mathrm{Dom}\,(\bar\partial\bar\partial^*)$ and $\mathrm{Range}\,(I+\bar\partial^*\bar\partial)^{-1} \subset \mathrm{Dom}\,(\bar\partial^*\bar\partial)$. (1.3.9) also shows that $S = (I+\bar\partial^*\bar\partial)^{-1} - \bar\partial\bar\partial^*(I+\bar\partial\bar\partial^*)^{-1}$, so since $\bar\partial^2 = 0$, $\mathrm{Range}\,(S) \subset \mathrm{Dom}\,(\bar\partial^*\bar\partial)$ and $\bar\partial^*\bar\partial S = \bar\partial^*\bar\partial(I+\bar\partial^*\bar\partial)^{-1}$. (1.3.10) likewise shows that $S = (I+\bar\partial\bar\partial^*)^{-1} - \bar\partial^*\bar\partial(I+\bar\partial^*\bar\partial)^{-1}$, so since $(\bar\partial^*)^2 = 0$, $\mathrm{Range}\,(S) \subset \mathrm{Dom}\,(\bar\partial\bar\partial^*)$ and $\bar\partial\bar\partial^* S = \bar\partial\bar\partial^*(I+\bar\partial\bar\partial^*)^{-1}$. Thus finally we see that $\mathrm{Range}\,(S) \subset \mathrm{Dom}\,(F_1)$ and

$$F_1 S = \bar\partial\bar\partial^*(I+\bar\partial\bar\partial^*)^{-1} + \bar\partial^*\bar\partial(I+\bar\partial^*\bar\partial)^{-1} + (I+\bar\partial\bar\partial^*)^{-1} + (I+\bar\partial^*\bar\partial)^{-1} - I = I.$$

Since F_1 is injective (in fact $(F_1\phi, \phi) \geq \|\phi\|^2$), we are done. Q.E.D.

It is, in fact, true that $F_1 = F$. This follows essentially from a theorem of Friedrichs [10] asserting the identity of weak and strong extensions of differential operators; cf. also Hörmander [16]. We shall not present the proof since we shall be working exclusively with the operator F; in Chapter 3 we shall present an easy proof that $F_1 = F$ in the cases where we can prove regularity theorems for F.

Let $\mathcal{H}^{p,q} = \mathcal{N}(F_1 - I)$ where $\mathcal{N}$ denotes nullspace. Clearly $\mathcal{H}^{p,q} \supset \mathcal{N}(\bar\partial) \cap \mathcal{N}(\bar\partial^*)$; on the other hand, if $\phi \in \mathcal{H}^{p,q}$, then $0 = ((F_1-I)\phi,\phi) = (\bar\partial\phi, \bar\partial\phi) + (\bar\partial^*\phi, \bar\partial^*\phi)$, so $\phi \in \mathcal{N}(\bar\partial) \cap \mathcal{N}(\bar\partial^*)$. F_1-I being self-adjoint, we have $(\text{Range}\,(F_1-I))^c = (\mathcal{H}^{p,q})^\perp$ where c denotes closure; since more-Range $(\bar\partial) \perp$ Range $(\bar\partial^*)$, we have the *weak orthogonal decomposition*

$$H_0^{p,q} = (\text{Range}\,(\bar\partial\bar\partial^*))^c \oplus (\text{Range}\,(\bar\partial^*\bar\partial))^c \oplus \mathcal{H}^{p,q}\,.$$

Among our principal objectives will be to show that under certain conditions the ranges of $\bar\partial$ and $\bar\partial^*$ are closed and hence we obtain a strong orthogonal decomposition analogous to the Hodge decomposition for compact manifolds. This type of result in potential theory on compact manifolds dates back to Weyl [47].

CHAPTER II

THE MAIN THEOREM

1. *Statement of the theorem*

We are almost ready to formulate an existence and regularity theorem for the equation $F\phi = \alpha$. If the estimate

$$Q(\phi, \phi) \geq c\|\phi\|_1^2 \tag{2.1.1}$$

held for some constant $c > 0$ and all $\phi \in \mathfrak{D}^{p,q}$, our job would be relatively easy. The $\bar{\partial}$-Neumann problem would then be *coercive*, and such problems can be handled by well-known general techniques, cf. [1]. Indeed, it will become clear in §2.3 how the Main Theorem could be greatly sharpened and simplified if (2.1.1) held. However, (2.1.1) is in general false, and we must settle for something weaker. In order to motivate our substitute estimate for (2.1.1), we do some calculations for 1-forms in $\mathbf{C}^n$.

Let therefore M be a bounded region of $\mathbf{C}^n$ with smooth boundary defined, as usual, by an equation $r = 0$. If $\phi \in \Lambda^{0,1}(\bar{M})$, we have $\bar{\partial}\phi = \Sigma_{j<k}\left(\frac{\partial\phi_j}{\partial\bar{z}_k} - \frac{\partial\phi_k}{\partial\bar{z}_j}\right) d\bar{z}_j \wedge d\bar{z}_k$ and $\vartheta\phi = -2\Sigma_1^n \frac{\partial\phi_j}{\partial z_j}$, and $\phi \in \mathfrak{D}^{0,1}$ if and only if $\Sigma\phi_j \frac{\partial r}{\partial z_j} = 0$ on bM. Thus if $\phi \in \mathfrak{D}^{0,1}$,

$$\|\bar{\partial}\phi\|^2 = 4\Sigma_{j<k}\left\|\frac{\partial\phi_j}{\partial\bar{z}_k} - \frac{\partial\phi_k}{\partial\bar{z}_j}\right\|^2 = 4\Sigma_{j,k=1}^n\left\|\frac{\partial\phi_j}{\partial\bar{z}_k}\right\|^2 - 4\Sigma_{j,k=1}^n\left(\frac{\partial\phi_j}{\partial\bar{z}_k}, \frac{\partial\phi_k}{\partial\bar{z}_j}\right). \tag{2.1.2}$$

Now
$$\Sigma_{jk}\left(\frac{\partial\phi_j}{\partial\bar{z}_k}, \frac{\partial\phi_k}{\partial\bar{z}_j}\right) = -\Sigma_{jk}\left(\frac{\partial^2\phi_j}{\partial z_j\partial\bar{z}_k}, \phi_k\right) + \Sigma_{jk}\int_{bM}\frac{\partial r}{\partial z_j}\frac{\partial\phi_j}{\partial\bar{z}_k}\bar{\phi}_k$$
$$= \Sigma_{jk}\left(\frac{\partial\phi_j}{\partial z_j}, \frac{\partial\phi_k}{\partial z_k}\right) + \Sigma_{jk}\int_{bM}\frac{\partial r}{\partial z_j}\frac{\partial\phi_j}{\partial\bar{z}_k}\bar{\phi}_k - \Sigma_{jk}\int_{bM}\frac{\partial r}{\partial\bar{z}_k}\frac{\partial\phi_j}{\partial z_j}\bar{\phi}_k .$$

But $\Sigma_k \frac{\partial r}{\partial \bar z_k} \bar\phi_k = \left(\overline{\Sigma_k \frac{\partial r}{\partial z_k} \phi_k}\right) = 0$ on bM, so the last term vanishes. Moreover, $\Sigma_k \frac{\partial r}{\partial z_k} \phi_k = 0$ on bM implies that the tangential derivatives of $\Sigma_k \frac{\partial r}{\partial z_k} \phi_k$ also vanish on bM, and it also means that $\Sigma_k \bar\phi_k \frac{\partial}{\partial \bar z_k} = \left(\overline{\Sigma_k \phi_k \frac{\partial}{\partial z_k}}\right)$ *is* a tangential derivative on bM. Therefore $\Sigma_k \bar\phi_k \frac{\partial}{\partial \bar z_k} \left(\Sigma_j \phi_j \frac{\partial r}{\partial z_j}\right) = 0$ on bM, which yields $\Sigma_{jk} \frac{\partial^2 r}{\partial z_j \partial \bar z_k} \phi_j \bar\phi_k = - \Sigma_{jk} \frac{\partial r}{\partial z_j} \frac{\partial \phi_j}{\partial \bar z_k} \phi_k$ on bM. Thus we have

$$\sum_{jk} \left(\frac{\partial \phi_j}{\partial \bar z_k}, \frac{\partial \phi_k}{\partial \bar z_j}\right) = \sum_{jk} \left(\frac{\partial \phi_j}{\partial z_j}, \frac{\partial \phi_k}{\partial z_k}\right) - \sum_{jk} \int_{bM} \frac{\partial^2 r}{\partial z_j \partial \bar z_k} \phi_j \bar\phi_k .$$

But the first term on the right is just $\frac{1}{4}\|\vartheta\phi\|^2$, so finally, substituting in (2.1.2), we obtain

$$\|\bar\partial\phi\|^2 = - \|\vartheta\phi\|^2 + 4 \sum_{jk} \left\|\frac{\partial \phi_j}{\partial \bar z_k}\right\|^2 + 4 \sum_{jk} \int_{bM} \frac{\partial^2 r}{\partial z_j \partial \bar z_k} \phi_j \bar\phi_k ,$$

or

$$\text{(2.1.3)} \qquad Q(\phi, \phi) = 4 \sum_{j,k} \left\|\frac{\partial \phi_j}{\partial \bar z_k}\right\|^2 + 4 \sum_{jk} \int_{bM} \frac{\partial^2 r}{\partial z_j \partial \bar z_k} \phi_j \bar\phi_k + \|\phi\|^2 .$$

Let us interpret equation (2.1.3). For each $p \in bM$, set $V_p = \{a \in \mathbf{C}^n : \Sigma_1^n \frac{\partial r}{\partial z_j}(p) a_j = 0\}$. The *Levi form* at p is the quadratic form L_p on V_p defined by $L_p(a) = \Sigma \frac{\partial^2 r}{\partial z_j \partial \bar z_k} a_j \bar a_k$. M is said to be *strongly pseudoconvex at* p if L_p is positive definite, and *strongly pseudoconvex* if it is strongly pseudoconvex at each $p \in bM$. If we define the norm $E(\cdot)$ on $\Lambda^{0,1}(\bar M)$ by $E(\phi)^2 = \Sigma_{jk} \left\|\frac{\partial \phi_j}{\partial \bar z_k}\right\|^2 + \int_{bM} |\phi|^2 + \|\phi\|^2$, we then have:

(2.1.4) PROPOSITION. *For some* $c > 0$ *and all* $\phi \in \mathcal{D}^{0,1}$, $Q(\phi,\phi) \le cE(\phi)^2$, *and if* M *is strongly pseudoconvex, then there exists* $c' > 0$ *such that* $Q(\phi,\phi) \ge c'E(\phi)^2$.

(2.1.5) COROLLARY. *If* $M \subset \mathbb{C}^n$ *is strongly pseudoconvex, then* $\phi = 0$ *whenever* $\phi \in \mathfrak{D}^{0,1}$ *and* $\bar{\partial}\phi = \vartheta\phi = 0$.

Proof: From (2.1.3),

$$\sum_{jk} \left\|\frac{\partial \phi_j}{\partial \bar{z}_k}\right\|^2 + \int_{bM} |\phi|^2 \leq c'(\|\bar{\partial}\phi\|^2 + \|\vartheta\phi\|^2) = 0 .$$

Thus each ϕ_j is holomorphic on M and vanishes on bM, hence is zero. Q.E.D.

The estimate $Q(\phi, \phi) \geq c'E(\phi)^2$ is going to be our substitute for (2.1.1). Back on our general manifold M, we define the norm E as follows. Let $\omega_1, \ldots, \omega_n$ be a local orthonormal basis for $\Lambda^{1,0}(\bar{M})$ on the patch $U \subset M'$ and let $L_1, \ldots, L_n$ be the dual vector fields. Then if $\phi \in \Lambda^{p,q}(\bar{M})$ has support in U, we write $\phi = \Sigma_{IJ}\, \phi_{IJ}\, \omega^I \wedge \bar{\omega}^J$ and set $E_U(\phi)^2 = \Sigma_{IJk} \|\bar{L}_k \phi_{IJ}\|^2 + \int_{bM} |\phi|^2 + \|\phi\|^2$. We then fix a partition of unity $\{\rho_i\}$ subordinate to a covering by such patches $\{U_i\}$ and set $E(\phi)^2 = \Sigma\, E_{U_i}(\rho_i\phi)^2$. Clearly a different partition of unity will yield an equivalent norm.

(2.1.6) PROPOSITION. *For some* $c > 0$ *and all* $\phi \in \Lambda^{p,q}(\bar{M})$, $E(\phi)^2 \leq c\|\phi\|_1^2$.

Proof: We need only show that $\int_{bM} |\phi|^2 \leq c\|\phi\|_1^2$, and it suffices to prove this inequality for functions. Letting σ be a smooth $(2n-1)$-form on $\bar{M}$ such that $\sigma \mid bM$ is the volume form on bM, we have by Stokes' theorem,

$$\int_{bM} |u|^2 = \int_{bM} |u|^2\sigma = \int_M d(|u|^2\sigma)$$

$$\leq \left|\int_M du \wedge (\bar{u}\sigma)\right| + \left|\int_M u(d\bar{u}) \wedge \sigma\right| + \left|\int_M |u|^2\, d\sigma\right|$$

$$\leq c(\|u\|\, \|u\|_1 + \|u\|^2) \leq c\|u\|_1^2$$

by the Schwarz inequality. Q.E.D.

Remark. One can show that in fact E is strictly weaker than $\| \|_1$ on $\mathfrak{D}^{p,q}$ (and also, *a fortiori*, on $\Lambda^{p,q}(\bar{M})$).

We say the *basic estimate* holds in $\mathfrak{D}^{p,q}$ if for some $c > 0$, $E(\phi)^2 \leq c\,Q(\phi,\phi)$ for all $\phi \in \mathfrak{D}^{p,q}$.

We are now ready for the Main Theorem. Recall that by the construction of the operator F, for each $\alpha \in H_0^{p,q}$ there is a unique $\phi \in \tilde{\mathfrak{D}}^{p,q}$ satisfying $F\phi = \alpha$.

(2.1.7) MAIN THEOREM. *Suppose the basic estimate holds in* $\mathfrak{D}^{p,q}$. *Given* $\alpha \in H_0^{p,q}$, *let* $\phi \in \tilde{\mathfrak{D}}^{p,q}$ *be the unique solution of* $F\phi = \alpha$. *If* U *is a subregion of* $\bar{M}$ *and* $\alpha|U \in \Lambda^{p,q}(U)$, *then* $\phi|U \in \Lambda^{p,q}(U)$. *Moreover, in this case, if* ζ *and* ζ_1 *are smooth real functions with* $\operatorname{supp}\zeta \subset \operatorname{supp}\zeta_1 \subset U$ *and* $\zeta_1 = 1$ *on* $\operatorname{supp}\zeta$, *then:*

(1) *if* $U \cap bM = \emptyset$, *for each integer* $s \geq 0$ *there is a constant* c_s *(depending on* ζ *and* ζ_1 *but independent of* α*) such that* $\|\zeta\phi\|_{s+2}^2 \leq c_s(\|\zeta_1\alpha\|_s^2 + \|\alpha\|^2)$;

(2) *if* $U \cap bM \neq \emptyset$, *for each integer* $s \geq 0$ *there is a constant* c_s *(depending on* ζ *and* ζ_1 *but independent of* α*) such that* $\|\zeta\phi\|_{s+1}^2 \leq c_s(\|\zeta_1\alpha\|_s^2 + \|\alpha\|^2)$.

The remainder of this chapter will be devoted to the proof of the Main Theorem, which will be accomplished in several stages. The first step will be to prove regularity and *a priori* estimates in the interior of M (part (1) of the theorem). This follows from well-known facts about elliptic operators in general; see, e.g., [7]. However, we shall present proofs from scratch for our special case as a way of warming up to the study of regularity at the boundary. This more difficult problem will be attacked by adding some terms onto the form Q to obtain a form Q^δ $(0 \leq \delta \leq 1,\ Q^0 = Q)$ for which the coercive estimate $Q^\delta(\psi,\psi) \geq \delta\|\psi\|_1^2$ holds. We then prove regularity of solutions of the modified equation $Q^\delta(\phi^\delta,\psi) = (\alpha,\psi)$ for all $\psi \in \mathfrak{D}^{p,q}$ – again, by essentially well-known techniques. Finally, we derive *a priori* estimates for such solutions that hold uniformly in δ, which will enable us to conclude the smoothness of the original ϕ and the validity of these estimates for ϕ.

The proofs of the estimates are largely a sophisticated exercise in integration by parts. To facilitate matters, we recall at the outset two elementary facts and state some notations.

Fact 1. If D_1 and D_2 are differential operators of order k_1 and k_2 respectively, then $[D_1, D_2] = D_1D_2 - D_2D_1$ is a differential operator of order $k_1 + k_2 - 1$.

Fact 2. For any $\epsilon > 0$ there exists $K > 0$ such that for all positive numbers a and b, $ab \le \epsilon a^2 + Kb^2$. (In fact, we may take $K = \frac{1}{4\epsilon}$, since $\epsilon a^2 + \frac{1}{4\epsilon} b^2 - ab = (\sqrt{\epsilon}\, a - \frac{1}{2\sqrt{\epsilon}} b)^2 \geq 0$.)

We shall generally write this relation as $ab \le (sc)a^2 + (\ell c)b^2$ where (sc) stands for "small constant" and (ℓc) stands for "large constant," with the understanding that (sc) may be chosen as small as we please by taking (ℓc) sufficiently large.

If A and B are functions on a set S, we use the notations $A \lesssim B$ and $A = \mathcal{O}(B)$ interchangeably to mean that for some $c > 0, |A(\sigma)| \le c|B(\sigma)|$ for all $\sigma \in S$. If A and B also depend on other parameters, we shall say "$A \lesssim B$ uniformly for $\sigma \in S$" to indicate that the constant c is independent of σ, although perhaps not of the other parameters. Further, we write $A \sim B$ to mean that $A \lesssim B$ and $B \lesssim A$. These conventions will obviate the writing of many useless constants.

If $x_1, \ldots, x_{2n}$ is a system of local coordinates, we set $D^j = \frac{1}{i}\frac{\partial}{\partial x_j}$ and use the standard multi-index notation D^α for products of the D^j's.

The basic estimate does not enter the picture until §2.4. Therefore, in stating our preliminary theorems, we shall not assume that the basic estimate holds unless we explicitly say so.

2. *Estimates and regularity in the interior*

We begin with the fundamental fact of life about strongly elliptic operators.

(2.2.1) THEOREM (Gårding's Inequality). $\|\psi\|_1^2 \lesssim Q(\phi,\phi)$ *for all* $\psi \in \Lambda_0^{p,q}(M)$. (*Remember that* $\Lambda_0^{p,q}(M)$ *is in general not dense in* $\tilde{\mathfrak{D}}^{p,q}$.)

Proof: By using a partition of unity it suffices to prove the inequality for forms ψ supported in a patch on which there exists a local orthonormal basis $\omega_1,\dots,\omega_n$ for $\Lambda^{1,0}(M)$ with dual vector fields $L_1,\dots,L_n$. We may then write $\psi = \Sigma_{IJ}\, \psi_{IJ}\, \omega^I \wedge \bar\omega^J$. From formulas (1.2.3) and (1.2.4) it follows by some straightforward but messy algebraic calculations (cf. the proof of Lemma (3.2.3)) that

$$\Box\psi = \sum_{IJK} (L_k \bar L_k \psi_{IJ})\, \omega^I \wedge \bar\omega^J + \text{lower order terms}$$

$$= \sum_{IJK} (\bar L_k L_k \psi_{IJ})\, \omega^I \wedge \bar\omega^J + \text{lower order terms}\,.$$

(This is essentially the same computation that proves that $\sigma(\Box,\eta) = -|\Pi_{0,1}\eta|^2$.) Since all boundary terms vanish these formulas yield

$$(\Box\psi,\psi) = \sum (L_k \bar L_k \psi_{IJ}, \psi_{IJ}) + \mathcal{O}(\|\psi\|_1\, \|\psi\|)$$

$$= \sum (\bar L_k \psi_{IJ}, \bar L_k \psi_{IJ}) + \mathcal{O}(\|\psi\|_1\, \|\psi\|)\,,$$

and

$$(\Box\psi,\psi) = \sum (\bar L_k L_k \psi_{IJ}, \psi_{IJ}) + \mathcal{O}(\|\psi\|_1\, \|\psi\|)$$

$$= \sum (L_k \psi_{IJ}, L_k \psi_{IJ}) + \mathcal{O}(\|\psi\|_1\, \|\psi\|)\,.$$

Therefore,

$$Q(\psi,\psi) = ((\Box + I)\psi,\psi) = \tfrac{1}{2}\sum (\|L_k\psi_{IJ}\|^2 + \|\bar L_k \psi_{IJ}\|^2) + \|\psi\|^2 + \mathcal{O}(\|\psi\|_1\|\psi\|)\,.$$

But all first-order derivatives are linear combinations of the L_k's and $\overline{L}_k$'s, hence

$$\|\psi\|_1^2 \lesssim Q(\psi,\psi) + \mathcal{O}(\|\psi\|_1 \|\psi\|)$$
$$\lesssim Q(\psi,\psi) + (sc)\|\psi\|_1^2 + (\ell c)\|\psi\|^2 ,$$

which implies the desired result since $\|\psi\|^2 \leq Q(\psi,\psi)$. Q.E.D.

We now prove the interior *a priori* estimates for ϕ in terms of $F\phi = (\square + I)\phi$ when ϕ is smooth. For this discussion we fix subregions $V \subset \overline{V} \subset U \subset \overline{U} \subset M$ and a real $\mathcal{C}^\infty$ function ζ_1 supported in U with $\zeta_1 = 1$ on $\overline{V}$. Using a partition of unity if necessary, we assume U is a coordinate patch with coordinates $x_1, \ldots, x_{2n}$.

(2.2.2) LEMMA. *For each real* $\zeta \in \Lambda_0^{0,0}(V)$, $\|\zeta\phi\|_1^2 \lesssim \|\zeta_1 F\phi\|^2 + \|\phi\|^2$ *uniformly for* $\phi \in \mathrm{Dom}\,(F) \cap \Lambda^{p,q}(U)$.

Proof: By Theorem (2.2.1), it suffices to estimate $Q(\zeta\phi,\zeta\phi)$. Since all boundary terms in integration by parts vanish, we have

$$\begin{aligned}(\bar\partial\zeta\phi, \bar\partial\zeta\phi) &= (\zeta\bar\partial\phi, \bar\partial\zeta\phi) + \mathcal{O}(\|\phi\|\;\|\zeta\phi\|_1)\\ &= (\bar\partial\phi, \zeta\bar\partial\zeta\phi) + \mathcal{O}(\|\phi\|\;\|\zeta\phi\|_1)\\ &= (\bar\partial\phi, \bar\partial\zeta^2\phi) + (\bar\partial\phi, [\zeta,\bar\partial]\zeta\phi) + \mathcal{O}(\|\phi\|\;\|\zeta\phi\|_1)\\ &= (\bar\partial\phi, \bar\partial\zeta^2\phi) + (\phi, \vartheta[\zeta,\bar\partial]\zeta\phi) + \mathcal{O}(\|\phi\|\;\|\zeta\phi\|_1)\\ &= (\bar\partial\phi, \bar\partial\zeta^2\phi) + \mathcal{O}(\|\phi\|\;\|\zeta\phi\|_1) ,\end{aligned}$$

the error terms being estimated by the Schwarz inequality together with the fact that $[\zeta, \bar\partial]$ is of order zero. Likewise,

$$(\vartheta\zeta\phi, \vartheta\zeta\phi) = (\vartheta\phi, \vartheta\zeta^2\phi) + \mathcal{O}(\|\phi\|\;\|\zeta\phi\|_1) .$$

Therefore

$$\begin{aligned}Q(\zeta\phi,\zeta\phi) &= Q(\phi,\zeta^2\phi) + \mathcal{O}(\|\phi\|\,\|\zeta\phi\|_1)\\ &= (F\phi,\zeta^2\phi) + \mathcal{O}(\|\phi\|\,\|\zeta\phi\|_1)\\ &\le \|\zeta_1 F\phi\|^2 + (sc)\|\zeta\phi\|_1^2 + (\ell c)\|\phi\|^2\end{aligned}$$

since $\zeta_1 = 1$ on $\operatorname{supp}\zeta$ and $\|\zeta^2\phi\| \lesssim \|\phi\|$. This estimate combined with Theorem (2.2.1) completes the proof. Q.E.D.

(2.2.3) COROLLARY. *The result remains valid if ζ is a matrix of functions in $\Lambda_0^{0,0}(V)$ acting on the components of ϕ.*

Proof: The proof is exactly the same. Q.E.D.

(2.2.4) LEMMA. *With ζ,ϕ as in Lemma (2.2.2), let β be a multi-index of order k and let D^β act on forms componentwise. Then*

$$Q(D^\beta\zeta\phi, D^\beta\zeta\phi) = Q(\phi, \zeta(D^\beta)'D^\beta\zeta\phi) + \mathcal{O}(\|\zeta'\phi\|_k\,\|\zeta\phi\|_{k+1})$$

where $(D^\beta)'$ is the formal adjoint of D^β, and ζ' denotes various matrices of functions involving ζ and its derivatives (for example, $\zeta'\phi = [\bar\partial,\zeta]\phi$).

Proof: As the arguments in this lemma will be used several times hereafter, we present all the gory details. Proceeding as in the proof of Lemma (2.2.2),

$$\begin{aligned}(\bar\partial D^\beta\zeta\phi, \bar\partial D^\beta\zeta\phi) &= (D^\beta\bar\partial\zeta\phi, \bar\partial D^\beta\zeta\phi) + \mathcal{O}(\|\zeta\phi\|_k\,\|\zeta\phi\|_{k+1})\\ &= (D^\beta\zeta\bar\partial\phi, \bar\partial D^\beta\zeta\phi) + \mathcal{O}(\|\zeta'\phi\|_k\,\|\zeta\phi\|_{k+1})\\ &= (\bar\partial\phi, \zeta(D^\beta)'\bar\partial D^\beta\zeta\phi) + \mathcal{O}(\|\zeta'\phi\|_k\,\|\zeta\phi\|_{k+1})\\ &= (\bar\partial\phi, \bar\partial\zeta(D^\beta)'D^\beta\zeta\phi) + (\bar\partial\phi, [\zeta(D^\beta)',\bar\partial]D^\beta\zeta\phi)\\ &\qquad + \mathcal{O}(\|\zeta'\phi\|_k\,\|\zeta\phi\|_{k+1})\,.\end{aligned}$$

But $(\bar{\partial}\phi, [\zeta(D^\beta)', \bar{\partial}]D^\beta\zeta\phi) = ([\vartheta, D^\beta\zeta]\bar{\partial}\phi, D^\beta\zeta\phi)$, and

$$[\vartheta, D^\beta\zeta]\bar{\partial}\phi = [\vartheta, D^\beta]\zeta\bar{\partial}\phi - D^\beta[\vartheta, \zeta]\bar{\partial}\phi$$

$$= [\vartheta, D^\beta]\bar{\partial}\zeta\phi + [\vartheta, D^\beta][\zeta, \bar{\partial}]\phi - D^\beta[[\vartheta, \zeta], \bar{\partial}]\phi - D^\beta\bar{\partial}[\vartheta, \zeta]\phi\ .$$

Taking the scalar product with $D^\beta\zeta\phi$, the first term is $\mathcal{O}(\|\zeta\phi\|_{k+1}\|\zeta\phi\|_k)$; the second and third terms are $\mathcal{O}(\|\zeta'\phi\|_k\|\zeta\phi\|_k) = \mathcal{O}(\|\zeta'\phi\|_k\|\zeta\phi\|_{k+1})$; and for the last we use the generalized Schwarz inequality (cf. Appendix, §1) to conclude that

$$(D^\beta\bar{\partial}[\vartheta, \zeta]\phi, D^\beta\zeta\phi) \lesssim \|D^\beta\bar{\partial}[\vartheta, \zeta]\phi\|_{-1}\|D^\beta\zeta\phi\|_1 \lesssim \|\zeta'\phi\|_k\|\zeta\phi\|_{k+1}\ .$$

After performing the same calculations for $(\vartheta D^\beta\zeta\phi, \vartheta D^\beta\zeta\phi)$ and adding, we are done. Q.E.D.

(2.2.5) THEOREM. *For each real* $\zeta \in \Lambda_0^{0,0}(V)$ *and each positive integer* s, $\|\zeta\phi\|_{s+2}^2 \lesssim \|\zeta_1 F\phi\|_s^2 + \|\phi\|^2$ *uniformly for* $\phi \in \mathrm{Dom}\,(F) \cap \Lambda^{p,q}(U)$.

Proof: For $s = 0$ we have $\|\zeta\phi\|_2^2 \sim \Sigma_j \|D^j\zeta\phi\|_1^2 + \|\zeta\phi\|_1^2 \lesssim \Sigma_j\|D^j\zeta\phi\|_1^2 + \|\zeta_1 F\phi\|^2 + \|\phi\|^2$ by Lemma (2.2.2). By Theorem (2.2.1) and Lemma (2.2.4),

$$\|D^j\zeta\phi\|_1^2 \lesssim Q(D^j\zeta\phi, D^j\zeta\phi)$$

$$= Q(\phi, \zeta(D^j)'D^j\zeta\phi) + \mathcal{O}(\|\zeta'\phi\|_1\|\zeta\phi\|_2)$$

$$= (F\phi, \zeta(D^j)'D^j\zeta\phi) + \mathcal{O}(\|\zeta'\phi\|_1\|\zeta\phi\|_2)$$

$$= (\zeta_1 F\phi, \zeta(D^j)'D^j\zeta\phi) + \mathcal{O}(\|\zeta'\phi\|_1\|\zeta\phi\|_2)$$

$$\leq (\ell c)\|\zeta_1 F\phi\|^2 + (sc)\|\zeta\phi\|_2^2 + (\ell c)\|\zeta'\phi\|_1^2 + (sc)\|\zeta\phi\|_2^2\ .$$

By Corollary (2.2.3), $\|\zeta'\phi\|_1^2 \lesssim \|\zeta_1 F\phi\|^2 + \|\phi\|^2$. Therefore, summing over j, we see that $\|\zeta\phi\|_2^2 \lesssim \|\zeta_1 F\phi\|^2 + \|\phi\|^2 + (sc)\|\zeta\phi\|_2^2$, which proves the theorem for $s = 0$.

By induction, suppose the theorem true for $s-1$. Then $\|\zeta\phi\|_{s+2}^2 \sim \Sigma_{|\beta|=s+1}\|D^\beta\zeta\phi\|_1^2 + \|\zeta\phi\|_{s+1}^2 \lesssim \Sigma_{|\beta|=s+1}\|D^\beta\zeta\phi\|_1^2 + \|\zeta_1 F\phi\|_{s-1}^2 + \|\phi\|^2$.

As above, by Theorem (2.2.1) and Lemma (2.2.4) we have

$$\begin{aligned}\|D^\beta\zeta\phi\|_1^2 &\lesssim Q(D^\beta\zeta\phi, D^\beta\zeta\phi)\\ &= Q(\phi, \zeta(D^\beta)' D^\beta\zeta\phi) + \mathcal{O}(\|\zeta'\phi\|_{s+1}\|\zeta\phi\|_{s+2})\\ &= (F\phi, \zeta(D^\beta)' D^\beta\zeta\phi) + \mathcal{O}(\|\zeta'\phi\|_{s+1}\|\zeta\phi\|_{s+2})\\ &\le |(\zeta_1 F\phi, \zeta(D^\beta)' D^\beta\zeta\phi)| + (\ell c)\|\zeta'\phi\|_{s+1}^2 + (sc)\|\zeta\phi\|_{s+2}^2 .\end{aligned}$$

By the generalized Schwarz inequality (cf. Appendix, §1),

$$\begin{aligned}|(\zeta_1 F\phi, \zeta(D^\beta)' D^\beta\zeta\phi)| &\le \|\zeta_1 F\phi\|_s \, \|\zeta(D^\beta)' D^\beta\zeta\phi\|_{-s}\\ &\le (\ell c)\|\zeta_1 F\phi\|_s^2 + (sc)\|\zeta\phi\|_{s+2}^2 .\end{aligned}$$

Finally, by inductive hypothesis, $\|\zeta'\phi\|_{s+1}^2 \lesssim \|\zeta_1 F\phi\|_{s-1}^2 + \|\phi\|^2 \lesssim \|\zeta_1 F\phi\|_s^2 + \|\phi\|^2$. Putting these facts together, we are done. Q.E.D.

Having shown that the derivatives of a smooth form ϕ can be estimated in terms of the derivatives of $F\phi$, we now show that ϕ is smooth inside M wherever $F\phi$ is. As a first step, we show that elements of $\tilde{\mathfrak{D}}^{p,q}$ are locally in $H_1^{p,q}$ inside M.

(2.2.6) LEMMA. *For each real* $\zeta \in \Lambda^{0,0}(\bar M)$, $Q(\zeta\phi, \zeta\phi) \lesssim Q(\phi, \phi)$ *uniformly for* $\phi \in \mathfrak{D}^{p,q}$.

Proof:
$$\begin{aligned}(\bar\partial\zeta\phi, \bar\partial\zeta\phi) &= (\zeta\bar\partial\phi, \zeta\bar\partial\phi) + ([\bar\partial, \zeta]\phi, [\bar\partial, \zeta]\phi) + 2\,\mathrm{Re}(\zeta\bar\partial\phi, [\bar\partial, \zeta]\phi)\\ &\lesssim \|\zeta\bar\partial\phi\|^2 + \|\zeta'\phi\|^2 + \|\zeta\bar\partial\phi\| \, \|\zeta'\phi\|\\ &\lesssim \|\bar\partial\phi\|^2 + \|\phi\|^2 .\end{aligned}$$

Likewise $(\vartheta\zeta\phi, \vartheta\zeta\phi) \lesssim \|\vartheta\phi\|^2 + \|\phi\|^2$, so $Q(\zeta\phi, \zeta\phi) \lesssim \|\bar\partial\phi\|^2 + \|\vartheta\phi\|^2 + \|\phi\|^2 = Q(\phi, \phi)$. Q.E.D.

(2.2.7) LEMMA. *For each real* $\zeta \in \Lambda_0^{0,0}(M)$, $\zeta\phi \in H_1^{p,q}$ *for all* $\phi \in \tilde{\mathfrak{D}}^{p,q}$.

Proof: Choose a sequence $\{\phi_n\} \subset \mathfrak{D}^{p,q}$ which is Q-convergent to ϕ. Then by Theorem (2.2.1) and Lemma (2.2.6),

$$\|\zeta(\phi_n - \phi_m)\|_1^2 \lesssim Q(\zeta(\phi_n - \phi_m), \zeta(\phi_n - \phi_m)) \lesssim Q(\phi_n - \phi_m, \phi_n - \phi_m) \to 0$$

as $m, n \to \infty$. Hence $\{\zeta\phi_n\}$ is a Cauchy sequence in $H_1^{p,q}$, and its limit is $\zeta\phi$. Q.E.D.

We now prove the interior regularity theorem by means of the technique of difference operators developed by Nirenberg [36]. Assuming as usual that we are working in a coordinate patch inside M, and (without loss of generality) that the range of the coordinate mapping is all of R^{2n}, we define the operator Δ_h^j on functions by

$$\Delta_h^j u(x) = \frac{1}{2ih}[u(x_1, \ldots, x_{j-1}, x_j + h, x_{j+1}, \ldots, x_{2n}) - u(x_1, \ldots, x_{j-1}, x_j - h, x_{j+1}, \ldots, x_{2n})]$$

Δ_h^j is then defined on forms by componentwise action. Further, if β is a multi-index and $H = (h_{11}, \ldots, h_{1\beta_1}, \ldots, h_{n1}, \ldots, h_{n\beta_n})$, we set $\Delta_H^\beta = \Pi_{j=1}^n \Pi_{k=1}^{\beta_j} \Delta_{h_{jk}}^j$. The facts we need about Δ_H^β are summarized in the following lemma.

(2.2.8) LEMMA. *Let ϕ and ψ be forms supported in a fixed compact set, let* D *be a first-order differential operator, and let $|\beta| = s$. Then:*

(1) *If $\phi \in H_s^{p,q}$, $\|\Delta_H^\beta \phi\| \lesssim \|\phi\|_s$ uniformly as $H \to 0$ (as a vector in R^s).*

(2) *If $\phi \in H_s^{p,q}$, $\|[D, \Delta_H^\beta]\phi\| \lesssim \|\phi\|_s$ uniformly as $H \to 0$.*

(3) *If $\phi \in H_{s-1}^{p,q}$, $(\Delta_H^\beta \phi, \psi) = (\phi, \Delta_H^\beta \psi) + \mathcal{O}(\|\phi\|_{s-1}\|\psi\|)$ uniformly as $H \to 0$.*

(4) *If $\phi \in H_{s-1}^{p,q}$ and $\psi \in H_1^{p,q}$, $|(\phi, [\Delta_H^\beta, D]\psi)| \lesssim \|\phi\|_{s-1}\|\psi\|_1$ uniformly as $H \to 0$.*

(5) *If $\phi \in H_t^{p,q}$ and $\|\Delta_H^\beta \phi\|_t$ is bounded as $H \to 0$ then $D^\beta \phi \in H_t^{p,q}$.*

This lemma will be proved in the Appendix, §4.

(2.2.9) THEOREM. *Let* U, V *be regions with* $V \subset \bar{V} \subset U \subset \bar{U} \subset M$, *and let* ζ_1 *be a real* $\mathcal{C}^\infty$ *function supported in* U *with* $\zeta_1 = 1$ *on* V. *If* $\phi \in \mathrm{Dom}(F)$ *and* $\zeta_1 F\phi \in H_s^{p,q}$ *for some* $s \geq 0$, *then* $\zeta\phi \in H_{s+2}^{p,q}$ *for any real* $\zeta \in \Lambda_0^{0,0}(V)$.

Proof: We know $\zeta'\phi \in H_1^{p,q}$ for all ζ' supported in V by Lemma (2.2.7), so by (5) of Lemma (2.2.8), to prove the theorem for $s = 0$ it suffices to show $\|\Delta_H^\beta \zeta\phi\|_1^2$ is bounded as $H \to 0$ for $|\beta| = 1$. Continuing inductively, if the theorem is true for $s-1$ and $\zeta_1 F\phi \in H_s^{p,q} \subset H_{s-1}^{p,q}$, then $\zeta'\phi \in H_{s+1}^{p,q}$ for all ζ' supported in V and it suffices to show $\|\Delta_H^\beta \zeta\phi\|_1^2$ is bounded as $H \to 0$ for $|\beta| = s+1$. By Theorem (2.2.1), therefore, we are reduced to proving for all $s \geq 1$ that if $\zeta'\phi \in H_s^{p,q}$ for all ζ' supported in V and $\zeta_1 F\phi \in H_{s-1}^{p,q}$, then $Q(\Delta_H^\beta \zeta\phi, \Delta_H^\beta \zeta\phi)$ is bounded as $H \to 0$ for $|\beta| = s$. Using (1)-(4) of Lemma (2.2.8), we have

$$\begin{aligned}
(\bar\partial\Delta_H^\beta\zeta\phi, \bar\partial\Delta_H^\beta\zeta\phi) &= (\Delta_H^\beta\bar\partial\zeta\phi, \bar\partial\Delta_H^\beta\zeta\phi) + \mathcal{O}(\|\zeta\phi\|_s \|\Delta_H^\beta\zeta\phi\|_1) \\
&= (\Delta_H^\beta\zeta\bar\partial\phi, \bar\partial\Delta_H^\beta\zeta\phi) + \mathcal{O}(\|\zeta'\phi\|_s \|\Delta_H^\beta\zeta\phi\|_1) \\
&= (\zeta\bar\partial\phi, \Delta_H^\beta\bar\partial\Delta_H^\beta\zeta\phi) + \mathcal{O}(\|\zeta'\phi\|_s \|\Delta_H^\beta\zeta\phi\|_1) \\
&= (\zeta\bar\partial\phi, \bar\partial\Delta_H^\beta\Delta_H^\beta\zeta\phi) + \mathcal{O}(\|\zeta'\phi\|_s \|\Delta_H^\beta\zeta\phi\|_1) \\
&= (\bar\partial\phi, \bar\partial\zeta\Delta_H^\beta\Delta_H^\beta\zeta\phi) + (\bar\partial\phi, [\zeta,\bar\partial]\Delta_H^\beta\Delta_H^\beta\zeta\phi) + \mathcal{O}(\|\zeta'\phi\|_s \|\Delta_H^\beta\zeta\phi\|_1)\,.
\end{aligned}$$

We have used the fact that $\|\zeta\bar\partial\phi\|_{s-1}^2 \lesssim \|\bar\partial\zeta\phi\|_{s-1}^2 + \|\zeta'\phi\|_{s-1}^2 \lesssim \|\zeta'\phi\|_s^2$. Now $[\zeta, \bar\partial]$ is a matrix of functions, so it can be moved to the other side of the scalar product along with $\Delta_{H'}^{\beta'}$ where $\Delta_{H'}^{\beta'}\Delta_h^j = \Delta_H^\beta$ for some j, h to yield

$$\begin{aligned}
(\bar\partial\phi, [\zeta,\bar\partial]\Delta_H^\beta\Delta_H^\beta\zeta\phi) &= (\Delta_{H'}^{\beta'}\zeta'\bar\partial\phi, \Delta_h^j\Delta_H^\beta\zeta\phi) + \mathcal{O}(\|\zeta'\bar\partial\phi\|_{s-2} \|\Delta_h^j\Delta_H^\beta\zeta\phi\|) \\
&= \mathcal{O}(\|\zeta'\phi\|_s \|\Delta_H^\beta\zeta\phi\|_1)\,.
\end{aligned}$$

After performing the same calculation for $(\vartheta\Delta_H^\beta\zeta\phi, \vartheta\Delta_H^\beta\zeta\phi)$, we have

$$\begin{aligned}Q(\Delta_H^\beta\zeta\phi, \Delta_H^\beta\zeta\phi) &= Q(\phi, \zeta\Delta_H^\beta\Delta_H^\beta\zeta\phi) + \mathcal{O}(\|\zeta'\phi\|_s\,\|\Delta_H^\beta\zeta\phi\|_1)\\ &= (F\phi, \zeta\Delta_H^\beta\Delta_H^\beta\zeta\phi) + \mathcal{O}(\|\zeta'\phi\|_s\,\|\Delta_H^\beta\zeta\phi\|_1)\\ &= (\Delta_{H'}^{\beta'}\zeta\zeta_1 F\phi, \Delta_h^j\Delta_H^\beta\zeta\phi) + \mathcal{O}(\|\zeta'\phi\|_s\,\|\Delta_H^\beta\zeta\phi\|_1\\ &\qquad + \|\zeta_1 F\phi\|_{s-2}\|\Delta_h^j\Delta_H^\beta\zeta\phi\|)\\ &= \mathcal{O}(\|\zeta_1 F\phi\|_{s-1}\,\|\Delta_H^\beta\zeta\phi\|_1 + \|\zeta'\phi\|_s\,\|\Delta_H^\beta\zeta\phi\|_1)\\ &\le (\ell c)\|\zeta_1 F\phi\|_{s-1}^2 + (\ell c)\|\zeta'\phi\|_s^2 + (sc)\|\Delta_H^\beta\zeta\phi\|_1^2\,.\end{aligned}$$

As this estimate holds uniformly as $H \to 0$, we are done. Q.E.D.

(2.2.10) COROLLARY. *If* $\phi \in \mathrm{Dom}\,(F)$ *and* $\zeta_1 F\phi \in \Lambda_0^{p,q}(M)$ *then* $\zeta\phi \in \Lambda_0^{p,q}(M)$.

Proof: This is an immediate consequence of Theorem (2.2.9) and the Sobolev lemma (cf. Appendix, §1.)

Remark. The results of this section do not depend in any way on the boundary conditions. Indeed, in the *a priori* estimates we could have dropped the assumption that $\phi \in \mathrm{Dom}\,(F)$ and replaced $Q(\phi, \psi)$ by $((\square + I)\phi, \psi)$ without any change. In the proof of regularity we used the fact that $\phi \in \tilde{\mathfrak{D}}^{p,q}$ only in order to approximate it by smooth forms in Lemma (2.2.7). We could have accomplished this by invoking Friedrichs' theorem [10], or we could have avoided it entirely with some additional work, cf. [7] or [36].

3. *Elliptic regularization*

The crucial fact which made all the arguments in §2.2 work was Gårding's inequality (2.2.1), and the crucial fact which makes life difficult at the boundary is that Gårding's inequality breaks down. To remedy this situation, we use the technique of elliptic regularization developed by Kohn and Nirenberg [27] which consists in adding an extra term onto

Q so that Gårding's inequality will hold, proving regularity for solutions of the modified problem, and then obtaining estimates valid uniformly as the extra term goes to zero which will enable us to conclude the regularity of the original solution. (This method was inspired by the idea of adding a "viscosity term" to regularize the equations of non-viscous fluid flow in hydrodynamics.)

Specifically, we define the form $Q^\delta (0<\delta\leq 1)$ by

$$Q^\delta(\phi,\psi) = Q(\phi,\psi) + \delta \sum_{i\epsilon I} \sum_{j=1}^{2n} (D^j\rho_i\phi, D^j\rho_i\psi)$$

where $\{\rho_i\}_{i\epsilon I}$ is a partition of unity subordinate to a coordinate covering and $\phi,\psi \,\epsilon\, \mathcal{D}^{p,q}$. We then extend Q^δ by continuity to $\tilde{\mathcal{D}}^{p,q}_\delta$, the completion of $\mathcal{D}^{p,q}$ under Q^δ.

(2.3.1) LEMMA. *$\tilde{\mathcal{D}}^{p,q}_\delta$ is independent of δ for $\delta>0$ and is contained in $\tilde{\mathcal{D}}^{p,q} \cap H^{p,q}_1$.*

Proof: Since $Q^\delta(\phi,\phi) \sim Q(\phi,\phi) + \delta\|\phi\|_1^2$, a sequence in $\mathcal{D}^{p,q}$ is Cauchy with respect to Q^δ if and only if it is Cauchy with respect to Q and $\| \;\|_1$. Q.E.D.

As before, the Friedrichs construction (1.3.3) associates to Q^δ an operator F^δ which is a self-adjoint extension of the differential operator $\Box + I + \Sigma_{ij}\, \rho_i (D^j)' D^j \rho_i$ by the argument in Proposition (1.3.5). Since the added term involves all first-order derivatives, F^δ is elliptic, and we clearly have the Gårding-type estimate $\delta\|\phi\|_1^2 \lesssim Q^\delta(\phi,\phi)$ for all $\phi \,\epsilon\, \tilde{\mathcal{D}}^{p,q}_\delta$.

A word of explanation about this construction. It could happen that the differential operator of which F^δ is a self-adjoint form is actually a constant multiple of $\Box + I$. In this case we have not essentially changed the operator under consideration; what has been changed is the free boundary condition, cf. Proposition (1.3.5). The new boundary conditions define a *coercive* problem, and this is what makes the method work.

Since $Q^{\delta}(\phi,\phi) \geq Q(\phi,\phi)$, the solution ϕ^{δ} of the equation $F^{\delta}\phi^{\delta} = a$ is unique, and the arguments of §2.2 go through without change to show that ϕ^{δ} is smooth inside M wherever a is and that the derivatives of ϕ^{δ} inside M can be estimated in terms of the derivatives of a inside M. But now we can prove similar theorems at the boundary.

A *special boundary chart* is a chart intersecting bM on which we have chosen coordinates $t_1, \ldots, t_{2n-1}, r$, where r is the function defining bM and $\{t_1, \ldots, t_{2n-1}\}_{r=0}$ are a coordinate system on bM, and a local orthonormal basis $\omega_1, \ldots, \omega_n$ for $\Lambda^{1,0}(\overline{M})$ such that $\omega_n = \sqrt{2}\,\partial r$. (The technique of using special boundary charts in this context was introduced by Ash [5].)

(2.3.2) LEMMA. *If* $\phi = \Sigma_{IJ}\, \phi_{IJ}\, \omega^I \wedge \overline{\omega}^J$ *in a special boundary chart, then* $\phi \in \mathfrak{D}^{p,q}$ *if and only if* $\phi \in \Lambda^{p,q}(\overline{M})$ *and* $\phi_{IJ} = 0$ *on* bM *whenever* $n \in J$.

Proof: Formula (1.2.4) implies that for $\phi \in \Lambda^{p,q}(\overline{M})$,

$$\sigma(\vartheta, dr)\phi = \frac{(-1)^p}{\sqrt{2}} \sum_{IHJ} \epsilon^{J}_{nH}\, \phi_{IJ}\, \omega^I \wedge \overline{\omega}^H ,$$

which, together with Proposition (1.3.2), proves the lemma. Q.E.D.

In a special boundary chart we define $D^j_t = \frac{1}{i}\frac{\partial}{\partial t_j}$, $D_r = \frac{1}{i}\frac{\partial}{\partial r}$, $\Delta^j_h = \Delta^{t_j}_h$, $\Delta^{\beta}_H = \Pi^n_{j=1}\, \Pi^{\beta_j}_{k=1}\, \Delta^{t_j}_{h_{jk}}$. These operators act on forms componentwise *with respect to the special basis* $\omega_1, \ldots, \omega_n$. We then have:

(2.3.3) LEMMA. D^{β}_t *and* Δ^{β}_H *preserve* $\mathfrak{D}^{p,q}$, *and* Δ^{β}_H *preserves* $\tilde{\mathfrak{D}}^{p,q}_{\delta}$.

Proof: The first statement follows immediately from Lemma (2.3.2) since D^{β}_t and Δ^{β}_H are tangential along bM. Moreover, if $\{\phi_m\}$ is a sequence in $\mathfrak{D}^{p,q}$ converging to $\phi \in \tilde{\mathfrak{D}}^{p,q}_{\delta}$ with respect to Q^{δ}, then $\Delta^{\beta}_H \phi = Q^{\delta}\text{-}\lim \Delta^{\beta}_H \phi_m \in \tilde{\mathfrak{D}}^{p,q}_{\delta}$ since $\Delta^{\beta}_H \phi_m \in \mathfrak{D}^{p,q}$ and translations are Q^{δ}-continuous. Q.E.D.

It is also clear from Lemma (2.3.2) that multiplication by a smooth function preserves $\mathcal{D}^{p,q}$ and $\tilde{\mathcal{D}}^{p,q}_{\delta}$. With these facts in mind, we can prove the regularity theorem for F^{δ}.

(2.3.4) THEOREM. *Let* U *be a special boundary chart,* $V \subset \bar{V} \subset U$, *and* ζ_1 *a real smooth function supported in* U *with* $\zeta_1 = 1$ *on* $\bar{V}$. *If* $\phi \in \mathrm{Dom}(F^{\delta})$ *and* $\zeta_1 F^{\delta}\phi \in \Lambda^{p,q}(\bar{M})$, *then* $\zeta\phi \in \Lambda^{p,q}(\bar{M})$ *for any smooth* ζ *supported in* V.

Proof: By Lemma (2.3.1) we already know that $\zeta\phi \in H_1^{p,q}$. Proceeding inductively, we shall assume $\zeta\phi \in H_s^{p,q}$ and show that $\zeta\phi \in H_{s+1}^{p,q}$; the theorem will then follow from the Sobolev lemma (cf. Appendix, §2). Suppose then that $\zeta\phi \in H_s^{p,q}$. Lemma (2.2.8) remains true at the boundary (cf. Appendix, §4), and therefore, by virtue of Lemma (2.3.3), the proof of Theorem (2.2.9) goes through without change to show that $D_t^{\beta}\zeta\phi \in H_1^{p,q}$ for $|\beta| = s$. It remains only to show $D_t^{\beta} D_r^m \zeta\phi$ (or, equivalently, $\zeta D_t^{\beta} D_r^m \phi$) is in $H_0^{p,q}$ for $|\beta| + m = s+1$ and $m \geq 2$. We now use the fact that F^{δ} is elliptic, so that

$$F^{\delta}\phi = A_0 D_r^2 \phi + \sum_{j=1}^{2n-1} A_j D_t^j D_r \phi + \sum_{j,k=1}^{2n-1} A_{jk} D_t^j D_t^k \phi + B_0 D_r \phi + \\ + \sum_{j=1}^{2n-1} B_j D_t^j \phi + C\phi$$

where the A's, B's, and C's are smooth matrices and the A's are *invertible*. Hence

$$(2.3.5) \quad \zeta D_r^2 \phi = \zeta A_0^{-1}[\zeta_1 F^{\delta}\phi - \sum_{j=1}^{2n-1} A_j D_t^j D_r \phi - \sum_{j,k=1}^{2n-1} A_{jk} D_t^j D_t^k \phi \\ - B_0 D_r \phi - \sum_{j=1}^{2n-1} B_j D_j^t \phi - C\phi]$$

where we have slipped in a factor of ζ_1 in front of $F^{\delta}\phi$. Applying ζD_t^{β} $(|\beta| = s-1)$ to (2.3.5), we see that $\zeta D_t^{\beta} D_r^2 \phi$ is expressed in terms of derivatives of $\zeta_1 F^\delta \phi$ and derivatives of $\zeta\phi$ which we already know to be in $H_0^{p,q}$. Proceeding by induction on m, we apply $\zeta D_t^\beta D_r^{m-2}$ ($|\beta| + m = s+1$) to (2.3.5) and find that $\zeta D_t^\beta D_r^m \phi$ is expressed in terms of derivatives of $\zeta_1 F^\delta \phi$ and derivatives of $\zeta\phi$ of order $\leq s+1$ and r-order $\leq m-1$, which we have already handled. (To be precise, one must commute ζ through the differential operators, but this results only in lower-order errors.) Thus we finally obtain $D_t^\beta D_r^m \zeta\phi \in H_0^{p,q}$ for all $|\beta| + m = s+1$, so $\zeta\phi \in H_{s+1}^{p,q}$. Q.E.D.

Remark: We could also prove the *a priori* estimates $\|\zeta\phi\|_{s+2}^2 \lesssim \|\zeta_1 F^\delta \phi\|_s^2 + \|\phi\|^2$ under the conditions of Theorem (2.3.4). Indeed, we estimate the tangential derivatives of $\zeta\phi$ by the method of Theorem (2.2.5), using Lemma (2.3.3), and then use equation (2.3.5) to estimate the normal derivatives. However, we shall have no use for these estimates, since they depend strongly on the inequality $\|\phi\|_1^2 \leq \delta^{-1} Q^\delta(\phi,\phi)$, so the constants involved blow up as $\delta \to 0$. We therefore leave the details to the interested reader.

4. *Estimates at the boundary*

We now derive *a priori* estimates at the boundary for smooth solutions ϕ of the equation $F\phi = \alpha$, which will then provide uniform estimates for the solutions ϕ^δ of the equations $F^\delta \phi^\delta = \alpha$ as $\delta \to 0$. For this purpose it is necessary to assign a special prominence to the tangential derivatives along the boundary. We therefore define the *tangential Fourier transform* for smooth functions in a special boundary chart by

$$\tilde{u}(\tau, r) = (2\pi)^{-(2n-1)/2} \int_{R^{2n-1}} e^{-i\langle t,\tau\rangle} u(t,r)\, dt\ .$$

We define the operators Λ_t^s ($s \in R$, t meaning tangential) by

$$\Lambda_t^s u(t,r) = (2\pi)^{-(2n-1)/2} \int_{R^{2n-1}} e^{i\langle t,\tau\rangle} (1+|\tau|^2)^{s/2}\, \tilde{u}(\tau,r)\, d\tau$$

and then define the *tangential Sobolev norms* $|||\ \ |||_s$ by

$$|||u|||_s^2 = \|\Lambda_t^s u\|^2 = \int_{R^{2n-1}} \int_{-\infty}^{0} (1+|\tau|^2)^s |\tilde{u}(\tau, r)|^2 dr\, d\tau .$$

As usual, Λ_t^s and $|||\ \ |||_s$ are defined componentwise for forms. These norms therefore measure derivatives in the tangential directions; for a detailed discussion, see the Appendix, §3.

When we wish to consider derivatives in all directions at once in a special boundary chart, we shall employ the notation $D^j = D_t^j$ for $1 \le j \le 2n-1$ and $D^{2n} = D_r$. We also define

$$|||D\phi|||_s^2 = \sum_1^{2n} |||D^j\phi|||_s^2 + |||\phi|||_s^2 \sim |||\phi|||_{s+1}^2 + |||D_r\phi|||_s^2 .$$

Lastly, if U is a special boundary chart, we define $\Lambda_0^{p,q}(U \cap \bar{M})$ to be the subspace of $\Lambda^{p,q}(\bar{M})$ whose elements are supported in U (but do not necessarily vanish on bM).

Before proceeding to the *a priori* estimates we need some technical preliminaries. Let U be a special boundary chart and let ζ, ζ_1 be smooth functions supported in U with $\zeta_1 = 1$ on supp ζ. For $\phi \in \Lambda_0^{p,q}(U \cap \bar{M})$, set $A\phi = \zeta_1 \Lambda_t^k \zeta\phi$, where k is some number, and define A' by $(A'\phi, \psi) = (\phi, A\psi)$ for all $\psi \in \Lambda_0^{p,q}(U \cap \bar{M})$. Then A has the following properties.

(2.4.1) LEMMA. *For each* $s \in R$, *the following estimates hold uniformly for* $\phi \in \Lambda_0^{p,q}(U \cap \bar{M})$.

(1) $|||A\phi|||_s \lesssim |||\phi|||_{s+k}$, $|||A'\phi|||_s \lesssim |||\phi|||_{s+k}$;

(2) $|||(A - A')\phi|||_s \lesssim |||\phi|||_{s+k-1}$;

(3) *if* L *is any first-order differential operator, then* $|||[A, L]\phi|||_s \lesssim |||D\phi|||_{s+k-1}$, $|||[A-A', L]\phi|||_s \lesssim |||D\phi|||_{s+k-2}$, *and* $|||[A, [A, L]]\phi|||_s \lesssim |||D\phi|||_{s+2k-2}$;

(4) A *preserves* $\mathcal{D}^{p,q}$.

Proof: (1), (2), and (3) will be proved in the Appendix, §5. (4) follows immediately from Lemma (2.3.2). Q.E.D.

We need the following sharpened form of Lemma (2.2.4).

(2.4.2) LEMMA. $Q(A\phi, A\phi) - \operatorname{Re} Q(\phi, A'A\phi) = \mathcal{O}(|||D\phi|||_{k-1}^2)$ *uniformly for* $\phi \in \mathcal{D}^{p,q} \cap \Lambda_0^{p,q}(U \cap \bar{M})$.

Proof: This type of estimate is by now painfully familiar, but this one requires a new twist. To wit,

$$(\bar\partial A\phi, \bar\partial A\phi) - \operatorname{Re}(\bar\partial\phi, \bar\partial A'A\phi) = \tfrac{1}{2}\{2(\bar\partial A\phi, \bar\partial A\phi) - (\bar\partial\phi, \bar\partial A'A\phi) - (\bar\partial A'A\phi, \bar\partial\phi)\}$$

$$= -\tfrac{1}{2}\{([\bar\partial, A']A\phi, \bar\partial\phi) + (\bar\partial A\phi, [A, \bar\partial]\phi) +$$

$$+ (\bar\partial\phi, [\bar\partial, A]A\phi) + ([A, \bar\partial]\phi, \bar\partial A\phi)\}\ .$$

Let us consider the first and last terms on the right:

$$([\bar\partial, A']A\phi, \bar\partial\phi) + ([A, \bar\partial]\phi, \bar\partial A\phi)$$

$$= ([\bar\partial, A' - A]A\phi, \bar\partial\phi) + ([[\bar\partial, A], A]\phi, \bar\partial\phi) + ([\bar\partial, A]\phi, (A' - A)\bar\partial\phi) + ([\bar\partial, A]\phi, [A, \bar\partial]\phi)$$

$$\lesssim |||[\bar\partial, A' - A]A\phi|||_{1-k} |||\bar\partial\phi|||_{k-1} + |||[[\bar\partial, A], A]\phi|||_{1-k} |||\bar\partial\phi|||_{k-1} + |||D\phi|||_{k-1}^2$$

$$\lesssim |||DA\phi|||_{-1} |||D\phi|||_{k-1} + |||D\phi|||_{k-1}^2$$

where we have used properties (1)-(3) of A and, for the first two terms, the generalized Schwarz inequality (cf. Appendix, §3). But

$$|||DA\phi|||_{-1}^2 \lesssim \sum_1^{2n} |||AD^j\phi|||_{-1}^2 + \sum_1^{2n} |||[D^j, A]\phi|||_{-1}^2 + |||A\phi|||_{-1}^2$$

$$\lesssim |||D\phi|||_{k-1}^2 + |||D\phi|||_{k-2}^2 + |||\phi|||_{k-1}^2$$

$$\lesssim |||D\phi|||_{k-1}^2\ ,$$

which completes the estimate of the first and last terms.

The second and third terms are estimated by the same procedure. Finally, applying this argument to $(\vartheta A\phi, \vartheta A\phi) - \mathrm{Re}(\vartheta\phi, \vartheta A'A\phi)$ and adding, we are done. Q.E.D.

In case $k = 0$ we can say more.

(2.4.3) LEMMA. $Q(\zeta\phi, \zeta\phi) - \mathrm{Re}\, Q(\phi, \zeta^2\phi) = \mathcal{O}(\|\phi\|^2)$ *uniformly for* $\phi \in \mathfrak{D}^{p,q}$.

Proof: Letting A = multiplication by ζ, we follow through the proof of Lemma (2.4.2) up to the point where we estimate $([\bar{\partial}, A']A\phi, \bar{\partial}\phi) + ([A, \bar{\partial}]\phi, \bar{\partial}A\phi)$. But now $A = A'$, and $[\partial, A]$ is multiplication by a matrix of functions so that $[[\bar{\partial}, A], A] = 0$. Hence we are left only with $([\bar{\partial}, A]\phi, [\bar{\partial}, A]\phi)$, which is $\mathcal{O}(\|\phi\|^2)$. Q.E.D.

Remember the basic estimate? We are ready to bring it into play via the following theorem.

(2.4.4) THEOREM. *For every* $p \in bM$ *there is a (small) special boundary chart* V *containing* p *such that* $|||D\phi|||^2_{-1/2} \lesssim E(\phi)^2$ *uniformly for* $\phi \in \Lambda_0^{p,q}(V \cap \bar{M})$.

It is convenient to prove instead the following more general theorem, from which Theorem (2.4.4) follows by taking $M_k = L_k$, $k = 1, \dots, n$.

(2.4.5) THEOREM. *Let* U *be a special boundary chart, and let* $M_1, \dots, M_N$ *be homogeneous first-order operators on* U, *say* $M_k = \Sigma a_{jk} D^j$, *such that there is no real* $\eta \neq 0 \in T^*U$ *with* $\sigma(M_k, \eta) = 0$ *for all* k. *Then for each* $p \in bM \cap U$ *there is a neighborhood* $V \subset U$ *of* p *such that* $\Sigma_1^{2n} |||D^j\phi|||^2_{-1/2} \lesssim \Sigma_1^N |||M_k\phi|||^2_{-1/2} + \int_{bM} |\phi|^2$ *for all* $\phi \in \Lambda_0^{p,q}(V \cap \bar{M})$.

Note that the estimate in Theorem (2.4.5) is strictly sharper than the one in Theorem (2.4.4), since $|||\ |||_{-1/2}$ is weaker than $\|\ \|$. Unfortunately, no one has yet found a way to use this additional information.

Proof of Theorem (2.4.5): Let $V \subset U$ be a neighborhood of p, and let ζ be a smooth function with $\zeta \leq 1$, $\zeta = 1$ on V, and $W = \operatorname{supp} \zeta \subset U$. Since the norms are defined componentwise, it suffices to prove the theorem for functions $u \in \Lambda_0^{0,0}(V \cap \overline{M})$ when V is sufficiently small. Let N_k be the operator M_k with coefficients frozen at p — that is, $N_k = \Sigma a_{jk}(p) D^j$ — and set $b_{jk}(x) = a_{jk}(x) - a_{jk}(p)$. Then for all $u \in \Lambda_0^{0,0}(V \cap \overline{M})$,

$$\begin{aligned}
|||(M_k - N_k)u|||_{-1/2} &\leq \sum_1^{2n} |||b_{jk} D^j u|||_{-1/2} = \sum |||\zeta b_{jk} D^j u|||_{-1/2} \\
&= \sum \|\Lambda_t^{-1/2} \zeta b_{jk} D^j u\| \\
&\leq \sum \|\zeta b_{jk} \Lambda_t^{-1/2} D^j u\| + \sum \|[\Lambda_t^{-1/2}, \zeta b_{jk}] D^j u\| \\
&\lesssim \sum (\sup_W |\zeta b_{jk}|)\, |||D^j u|||_{-1/2} + \sum |||D^j u|||_{-3/2},
\end{aligned}$$

where the last estimate depends on the fact that $[\Lambda_t^{-1/2}, \zeta b_{jk}]$ is of tangential order $-3/2$ (cf. Appendix, §5). Given any $\epsilon > 0$, we can choose W small enough so that $\sup_W |\zeta b_{jk}| < \epsilon$ since $b_{jk}(p) = 0$, and we can choose V small enough so that $|||D^j u|||_{-3/2} \leq \epsilon\, |||D^j u|||_{-1/2}$ (cf. Appendix, §3). Then we have $|||N_k u|||_{-1/2} \geq |||M_k u|||_{-1/2} - |||(M_k - N_k)u|||_{-1/2} \geq |||M_k u|||_{-1/2} - (\mathrm{sc}) \Sigma_1^{2n} |||D^j u|||_{-1/2}$, so it suffices to prove the theorem for the N_k's, which still satisfy the symbol condition. For this purpose we need some elementary facts about Fourier transforms, cf. [42].

First we prove that $\Sigma_1^{2n} |||D^j u|||_{-1/2}^2 \lesssim \Sigma_1^N |||N_k u|||_{-1/2}^2$ for functions u which vanish on bM, i.e., $u(t,0) = 0$. In this case we extend u to be zero outside $\overline{M}$; then $D_t^j u$ is continuous and $D_r u$ has only a jump discontinuity on bM, so these derivatives are square-integrable in V and we may apply the full Fourier transform to them. Letting ρ be the dual variable to r and setting $\xi = (\tau, \rho)$, we have

$$\begin{aligned}
\sum_1^N |||N_k u|||^2_{-1/2} &= \sum_1^N \|(1+|\tau|^2)^{-1/4}\widetilde{N_k u}(\tau, r)\|^2 \\
&= \sum_1^N \|(1+|\tau|^2)^{-1/4}\left(\sum_1^{2n-1} a_{jk}(p)\tau_j + a_{(2n)k}(p)\rho\right)\hat{u}(\xi)\|^2 \\
&= \sum_1^N \int_{R^{2n}} (1+|\tau|^2)^{-1/2}|\sigma(N_k,\xi)|^2|\hat{u}(\xi)|^2 d\xi \\
&\gtrsim \int_{R^{2n}} (1+|\tau|^2)^{-1/2}|\xi|^2|\hat{u}(\xi)|^2 d\xi \\
&= \int_{R^{2n}} (1+|\tau|^2)^{-1/2}\sum_1^{2n}|\widehat{D^j u}(\xi)|^2 d\xi \\
&= \int_{R^{2n-1}}\int_R (1+|\tau|^2)^{-1/2}\sum_1^{2n}|\widetilde{D^j u}(\tau, r)|^2 dr\, d\tau \\
&= \sum_1^{2n} |||D^j u|||^2_{-1/2}\,.
\end{aligned}$$

The second and sixth lines are applications of the Plancherel theorem, and the fourth line follows from our assumption on $\sigma(N_k, \xi)$.

Now for general u, we define $w(t, r)$ by

$$\tilde{w}(\tau, r) = \exp[(1+|\tau|^2)^{1/2} r]\,\tilde{u}(\tau, 0)$$

and set $v = u - w$. Since $u(t, 0) = w(t, 0)$, the preceding argument (which did not use any assumptions about compact support) applies to v, so

$$\begin{aligned}
\sum_1^{2n} |||D^j u|||^2_{-1/2} &\lesssim \sum_1^{2n} |||D^j v|||^2_{-1/2} + \sum_1^{2n} |||D^j w|||^2_{-1/2} \\
&\lesssim \sum_1^N |||N_k v|||^2_{-1/2} + \sum_1^{2n} |||D^j w|||^2_{-1/2} \\
&\lesssim \sum_1^N |||N_k u|||^2_{-1/2} + \sum_1^N |||N_k w|||^2_{-1/2} + \sum_1^{2n} |||D^j w|||^2_{-1/2} \\
&\lesssim \sum_1^N |||N_k u|||^2_{-1/2} + \sum_1^{2n} |||D_j w|||^2_{-1/2}
\end{aligned}$$

where in the last step we have expressed $N_k w$ in terms of the $D^j w$'s. It therefore suffices to show $|||D^j w|||^2_{-1/2} \lesssim \int_{bM} |u|^2$ for each j. For $j = 1, \ldots, 2n-1$, we have

$$|||D^j_t w|||^2_{-1/2} = \int_{R^{2n-1}} \int_{-\infty}^{0} (1+|\tau|^2)^{-1/2} |\tau_j|^2 \exp[2(1+|\tau|^2)^{1/2} r] |\tilde{u}(\tau,0)|^2 \, dr \, d\tau$$

$$\leq \int_{R^{2n-1}} \int_{-\infty}^{0} (1+|\tau|^2)^{1/2} \exp[2(1+|\tau|^2)^{1/2} r] |\tilde{u}(\tau,0)|^2 \, dr \, d\tau$$

$$= \int_{R^{2n-1}} \left[\int_{-\infty}^{0} e^{2s} ds \right] |\tilde{u}(\tau,0)|^2 \, d\tau$$

$$= \tfrac{1}{2} \int_{R^{2n-1}} |\tilde{u}(\tau,0)|^2 \, d\tau$$

$$\sim \int_{bM} |u|^2 \quad \text{(by the Plancherel theorem).}$$

For $j = 2n$, we have

$$|||D_r w|||^2_{-1/2} = \int_{R^{2n-1}} \int_{-\infty}^{0} (1+|\tau|^2)^{-1/2} |\widetilde{D_r w}(\tau, r)|^2 \, dr \, d\tau$$

$$= \int_{R^{2n-1}} \int_{-\infty}^{0} (1+|\tau|^2)^{-1/2} (1+|\tau|^2) \exp[2(1+|\tau|^2)^{1/2} r] |\tilde{u}(\tau,0)|^2 dr \, d\tau$$

$$= \int_{R^{2n-1}} \left[\int_{-\infty}^{0} e^{2s} ds \right] |\tilde{u}(\tau,0)|^2 \, d\tau$$

$$= \tfrac{1}{2} \int_{R^{2n-1}} |\tilde{u}(\tau,0)|^2 \, d\tau$$

$$\sim \int_{bM} |u|^2 . \qquad \text{Q.E.D.}$$

The heart of the *a priori* estimates is contained in the following lemma.

(2.4.6) LEMMA. *Suppose the basic estimate holds in* $\mathfrak{D}^{p,q}$. *Let* V *be a special boundary chart on which the conclusions of Theorem* (2.4.4) *hold, and let* $\{\zeta_k\}_1^\infty$ *be a sequence of real functions in* $\Lambda_0^{0,0}(V \cap \bar{M})$ *such that* $\zeta_k = 1$ *on* $\operatorname{supp} \zeta_{k+1}$. *Then for each positive integer* k, $|||D\zeta_k\phi|||^2_{(k-2)/2} \lesssim |||\zeta_1 F\phi|||^2_{(k-2)/2} + \|F\phi\|^2$ *uniformly for* $\phi \in \operatorname{Dom}(F) \cap \mathfrak{D}^{p,q}$.

Proof: Combining the basic estimate with Theorem (2.4.4), we have $|||D\psi|||^2_{-1/2} \lesssim Q(\psi,\psi)$ for all $\psi \in \mathfrak{D}^{p,q} \cap \Lambda_0^{p,q}(V \cap \bar{M})$. Using this estimate, we proceed by induction on k. For $k=1$, by Lemma (2.4.3) we have

$$
\begin{aligned}
|||D\zeta_1\phi|||^2_{-1/2} \lesssim Q(\zeta_1\phi,\zeta_1\phi) &\lesssim \operatorname{Re} Q(\phi,\zeta_1^2\phi) + \mathcal{O}(\|\phi\|^2) \\
&= \operatorname{Re}(F\phi,\zeta_1^2\phi) + \mathcal{O}(\|\phi\|^2) \\
&\lesssim \|F\phi\|\,\|\phi\| + \mathcal{O}(\|\phi\|^2) \\
&\lesssim \|F\phi\|^2
\end{aligned}
$$

since $\|\phi\| \le \|F\phi\|$.

Assume the lemma true for $k-1$ $(k>1)$; we shall prove it for k. Writing $\Lambda_t^{(k-1)/2} = \Lambda$ for short, we have $|||D\zeta_k\phi|||^2_{(k-2)/2} = |||D\Lambda\zeta_1\zeta_k\phi|||^2_{-1/2}$. Since D^j and Λ commute and $\zeta_k\zeta_{k-1} = \zeta_k$,

$$
\begin{aligned}
D^j\Lambda\zeta_1\zeta_k\phi &= D^j\zeta_1\Lambda\zeta_k\phi + D^j[\Lambda,\zeta_1]\zeta_k\phi \\
&= D^j\zeta_1\Lambda\zeta_k\phi + [\Lambda,[D^j,\zeta_1]]\zeta_k\zeta_{k-1}\phi + [\Lambda,\zeta_1][D^j,\zeta_k]\zeta_{k-1}\phi \\
&\quad + [\Lambda,\zeta_1]\zeta_k D^j\zeta_{k-1}\phi,
\end{aligned}
$$

and therefore

$$
\begin{aligned}
|||D\zeta_k\phi|||^2_{(k-2)/2} &\lesssim |||D\zeta_1\Lambda\zeta_k\phi|||^2_{-1/2} + |||\zeta_{k-1}\phi|||^2_{(k-3)/2} \\
&\quad + |||D\zeta_{k-1}\phi|||^2_{(k-3)/2} \\
&\lesssim |||D\zeta_1\Lambda\zeta_k\phi|||^2_{-1/2} + |||D\zeta_{k-1}\phi|||^2_{(k-3)/2},
\end{aligned}
\tag{2.4.7}
$$

where we have used the fact that the commutator of Λ with a function is an operator of tangential order $(k-3)/2$ (cf. Appendix, §5). Setting $A = \zeta_1 \Lambda \zeta_k$, we now apply Lemma (2.4.2):

$$
\begin{aligned}
|||DA\phi|||^2_{-1/2} = |||DA\zeta_{k-1}\phi|||^2_{-1/2} &\lesssim Q(A\zeta_{k-1}\phi, A\zeta_{k-1}\phi) \\
&= \operatorname{Re} Q(\zeta_{k-1}\phi, A'A\zeta_{k-1}\phi) + \mathcal{O}(|||D\zeta_{k-1}\phi|||^2_{(k-3)/2}) \\
&= \operatorname{Re} Q(\phi, A'A\phi) + \mathcal{O}(|||D\zeta_{k-1}\phi|||^2_{(k-3)/2}) \\
&= \operatorname{Re} (F\phi, A'A\phi) + \mathcal{O}(|||D\zeta_{k-1}\phi|||^2_{(k-3)/2}) \\
&= \operatorname{Re} (A\zeta_1 F\phi, A\phi) + \mathcal{O}(|||D\zeta_{k-1}\phi|||^2_{(k-3)/2}) \\
&\lesssim |||A\zeta_1 F\phi|||_{-1/2}\, |||A\phi|||_{1/2} + |||D\zeta_{k-1}\phi|||^2_{(k-3)/2} \\
&\lesssim |||\zeta_1 F\phi|||_{(k-2)/2}\, |||\zeta_k\phi|||_{k/2} + |||D\zeta_{k-1}\phi|||^2_{(k-3)/2} \\
&\lesssim (\ell c)|||\zeta_1 F\phi|||_{(k-2)/2} + (sc)|||D\zeta_k\phi|||^2_{(k-2)/2} \\
&\qquad + |||D\zeta_{k-1}\phi|||^2_{(k-3)/2}\,.
\end{aligned}
$$

Substituting this result in (2.4.7) and using the inductive hypothesis,

$$
\begin{aligned}
|||D\zeta_k\phi|||^2_{(k-2)/2} &\lesssim |||\zeta_1 F\phi|||^2_{(k-2)/2} + |||D\zeta_{k-1}\phi|||^2_{(k-3)/2} \\
&\lesssim |||\zeta_1 F\phi|||^2_{(k-2)/2} + |||\zeta_1 F\phi|||^2_{(k-3)/2} + \|F\phi\|^2 \\
&\lesssim |||\zeta_1 F\phi|||^2_{(k-2)/2} + \|F\phi\|^2\,. \qquad \text{Q.E.D.}
\end{aligned}
$$

(2.4.8) THEOREM. *Suppose the basic estimate holds in* $\mathcal{D}^{p,q}$. *Let* V *be a special boundary chart in which the conclusions of Theorem* (2.4.4) *hold. Let* $U \subset \overline{U} \subset V$, *and choose a real* $\zeta_1 \in \Lambda_0^{0,0}(V \cap \overline{M})$ *with* $\zeta_1 = 1$ *on* U. *Then for each real* $\zeta \in \Lambda_0^{0,0}(U \cap \overline{M})$ *and each positive integer* s, $\|\zeta\phi\|^2_{s+1} \lesssim \|\zeta_1 F\phi\|^2_s + \|F\phi\|^2$ *uniformly for* $\phi \in \operatorname{Dom}(F) \cap \mathcal{D}^{p,q}$.

Proof: We use induction on s. For $s=0$, we set $\zeta = \zeta_2$ and apply Lemma (2.4.6) with $k=2$ and $0 = \zeta_3 = \zeta_4 = \ldots$, obtaining

$$\|\zeta\phi\|_1^2 \sim \|D\zeta\phi\|^2 \lesssim \|\zeta_1 F\phi\| + \|F\phi\| .$$

Assume now the theorem true for s−1; then

$$\begin{aligned}\|\zeta\phi\|_{s+1}^2 &\sim \sum_{|\beta|=s+1} \|D^\beta\zeta\phi\|^2 + \|\zeta\phi\|_s^2 \\ &\lesssim \sum_{|\beta|=s+1} \|D^\beta\zeta\phi\|^2 + \|\zeta_1 F\phi\|_{s-1}^2 + \|F\phi\|^2 \\ &\lesssim \sum_{|\beta|=s+1} \|D^\beta\zeta\phi\|^2 + \|\zeta_1 F\phi\|_s^2 + \|F\phi\|^2 ,\end{aligned}$$

so we must estimate $\|D^\beta\zeta\phi\|^2$ for $|\beta| = s+1$. We interpolate a sequence of functions $\{\zeta_k\}_{k=2}^{2s+1}$ between ζ_1 and $\zeta = \zeta_{2s+2}$ such that $\zeta_k = 1$ on supp ζ_{k+1}, $k = 1, \ldots, 2s+1$, and apply Lemma (2.46) with $k = 2s+2$ (and $\zeta_j = 0$ for $j > 2s+2$):

$$\|D_t^\beta\zeta\phi\|^2 \lesssim |||D\zeta\phi|||_s^2 \lesssim |||\zeta_1 F\phi|||_s^2 + \|F\phi\|^2 \lesssim \|\zeta_1 F\phi\|_s^2 + \|F\phi\|^2$$

for $|\beta| = s+1$, and likewise for $|\beta| = s$,

$$\|D_t^\beta D_r\zeta\phi\|^2 \lesssim |||D\zeta\phi|||_s^2 \lesssim \|\zeta_1 F\phi\|_s^2 + \|F\phi\|^2 .$$

Thus it remains to estimate $D_t^\beta D_r^m\zeta\phi$ with $|\beta| + m = s + 1$, $m \geq 2$. But on applying $D_t^\beta D_r^{m-2}$ to the analogue of equation (2.3.5) for F^δ, we see – using induction on m – that $D_t^\beta D_r^m\zeta\phi$ is expressed in terms of derivatives of $\zeta_1 F\phi$ of order s−1 and derivatives of $\zeta\phi$ which we have already estimated. Q.E.D.

5. *Proof of the Main Theorem*

Having at last marshalled all our forces, we proceed without delay. Given $\alpha \in H_0^{p,q}$, let $\phi \in \tilde{\mathfrak{D}}^{p,q}$ be the unique solution of $F\phi = \alpha$. Let U be a subregion of M', $U \cap M \neq \emptyset$, and suppose $\alpha|U \in \Lambda^{p,q}(U \cap \bar{M})$. If $U \cap bM = \emptyset$, Corollary (2.2.10) shows that $\phi|U \in \Lambda^{p,q}(U)$ and Theorem (2.2.5) gives the norm estimate $\|\zeta\phi\|_{s+2}^2 \lesssim \|\zeta_1\alpha\|_s^2 + \|\alpha\|^2$ since $\|\phi\| \leq \|\alpha\|$. If U intersects bM, Theorem (2.4.8) provides the estimate

$\|\zeta\phi\|^2_{s+1} \lesssim \|\zeta_1\alpha\|^2_s + \|\alpha\|^2$ provided ϕ is smooth in $U \cap \overline{M}$. (We proved the estimate only for small U, but the general case now follows by a partition of unity argument.) It therefore remains only to show that $\phi|U \in \Lambda^{p,q}(U \cap \overline{M})$.

For each $\delta \in (0,1]$, let ϕ^δ be the solution of the regularized equation $F^\delta\phi^\delta = \alpha$. By Theorem (2.3.4), we know $\phi^\delta|U \in \Lambda^{p,q}(U \cap \overline{M})$. Also, since $Q^\delta(\psi,\psi) \geq Q(\psi,\psi)$, the arguments of §2.4 show that the estimates $\|\zeta\phi^\delta\|^2_{s+1} \lesssim \|\zeta_1\alpha\|^2_s + \|\alpha\|^2$ are valid. What is more, they hold *uniformly* in δ as $\delta \to 0$, since they depend essentially only on the estimate $|||D\zeta\phi^\delta|||^2_{-\frac{1}{2}} \lesssim Q(\zeta\phi^\delta,\zeta\phi^\delta) \sim Q^\delta(\zeta\phi^\delta,\zeta\phi^\delta) - \delta\|\zeta\phi^\delta\|^2_1$. The upshot is that $\{\zeta\phi^\delta\}_{0<\delta\leq 1}$ is a bounded set in $H^{p,q}_{s+1}$ for each s and therefore, by the Rellich lemma (cf. Appendix, §2), precompact in $H^{p,q}_s$ for each s. We can then extract for each s a subsequence $\{\zeta\phi^{\delta_n}\}$, $\delta_n \to 0$ as $n \to \infty$, which converges in $H^{p,q}_s$. If we can show that the limit of these subsequences is always $\zeta\phi$, we will have $\zeta\phi \in H^{p,q}_s$ for all s and all ζ and hence $\phi|U \in \Lambda^{p,q}(U \cap \overline{M})$ by the Sobolev lemma (cf. Appendix, §2), and we will be done.

It suffices to show that $\phi^\delta \to \phi$ in $H^{p,q}_0$ as $\delta \to 0$, for then if $\{\zeta\phi^{\delta_n}\}$ converges in any higher s-norm its limit must be $\zeta\phi$. Now the interior estimates for ϕ with $s = 0$ apply also to ϕ^δ uniformly in δ, so for any ζ supported in M we have $\zeta\phi^\delta \in H^{p,q}_2$ and

$$\|\zeta\phi^\delta\|_1 \lesssim \|\zeta\phi^\delta\|_2 \lesssim \|\alpha\| .$$

We know $\phi^\delta \in H^{p,q}_1$ by Lemma (2.3.1); applying a partition of unity together with the boundary estimate for $s = 0$, we see that

$$\|\phi^\delta\|_1 \lesssim \|\alpha\| \text{ uniformly as } \delta \to 0 . \tag{2.5.1}$$

(More precisely: this holds when α is globally smooth, hence in general since α can be approximated by smooth forms.) Next, for any $\psi \in \mathcal{D}^{p,q}$, by (2.5.1),

$$\begin{aligned} Q(\phi,\psi) = (\alpha,\psi) = Q^\delta(\phi^\delta,\psi) &= Q(\phi^\delta,\psi) + \mathcal{O}(\delta\|\phi^\delta\|_1\,\|\psi\|_1) \\ &= Q(\phi^\delta,\psi) + \mathcal{O}(\delta)\|\alpha\|\,\|\psi\|_1 . \end{aligned} \tag{2.5.2}$$

Writing the same equation for δ' and subtracting, we have $Q(\phi^\delta - \phi^{\delta'}, \psi) = \mathcal{O}(\delta - \delta')\,\|a\|\,\|\psi\|_1$. By Lemma (2.3.1) we can find a sequence $\{\psi_n\} \subset \mathcal{D}^{p,q}$ converging with respect to Q and $\|\ \|_1$ to $\phi^\delta - \phi^{\delta'}$, which yields (by (2.5.1) again)

$$Q(\phi^\delta - \phi^{\delta'}, \phi^\delta - \phi^{\delta'}) = \mathcal{O}(\delta - \delta')\,\|a\|\,\|\phi^\delta - \phi^{\delta'}\|_1 = \mathcal{O}(\delta - \delta')\,\|a\|^2$$

$$\to 0 \quad \text{as} \quad \delta, \delta' \to 0\,.$$

Thus $\{\phi^\delta\}$ converges in $\tilde{\mathcal{D}}^{p,q}$ as $\delta \to 0$, and equation (2.5.2) shows that its limit is ϕ. *A fortiori,* $\phi^\delta \to \phi$ in $H_0^{p,q}$. Q.E.D.

CHAPTER III

INTERPRETATION OF THE MAIN THEOREM

1. *Existence and regularity theorems for the $\bar{\partial}$ complex*

In this section we shall derive a number of basic properties of the $\bar{\partial}$ complex, some of which were foreshadowed in the discussion of Chapter I. All of these results are rather easy corollaries of the Main Theorem and its constituent parts. Throughout this section we shall always assume without further mention that the basic estimate holds in $\mathcal{D}^{p,q}$ unless the contrary is explicitly stated.

First we derive a few more properties of the operator F.

(3.1.1) PROPOSITION. *With α, ϕ, U, ζ_1, ζ as in the Main Theorem, let $k = 1$ or 2 according as $U \cap bM \neq \emptyset$ or $U \cap bM = \emptyset$. If $\alpha|U \in H_s^{p,q}(U)$ then $\zeta\phi \in H_{s+k}^{p,q}$ and $\|\zeta\phi\|_{s+k}^2 \lesssim \|\zeta_1\alpha\|_s^2 + \|\alpha\|^2$.*

Proof: Let ζ_0 be a smooth function supported in U with $\zeta_0 = 1$ on supp ζ_1. Pick sequences $\{\beta_n\}, \{\gamma_n\}$ of smooth forms with supp $\beta_n \subset$ supp ζ_0, supp $\gamma_n \subset$ supp $(1-\zeta_0)$, $\beta_n \to \zeta_0\alpha$ in $H_s^{p,q}$, and $\gamma_n \to (1-\zeta_0)\alpha$ in $H_0^{p,q}$. Then $\alpha_n = \beta_n + \gamma_n \to \alpha$ in $H_0^{p,q}$, and $\zeta_1\alpha_n \to \zeta_1\alpha$ in $H_s^{p,q}$. Let $\phi_n = F^{-1}\alpha_n$; since F^{-1} is bounded, $\phi_n \to \phi$ in $H_0^{p,q}$. By the Main Theorem,

$$\|\zeta(\phi_n - \phi_m)\|_{s+k} \lesssim \|\zeta_1(\alpha_n - \alpha_m)\|_s + \|\alpha_n - \alpha_m\| ,$$

from which it follows that $\zeta\phi = \lim \zeta\phi_n$ is in $H_{s+k}^{p,q}$ and

$$\|\zeta\phi\|_{s+k} \lesssim \|\zeta_1\alpha\|_s + \|\alpha\| . \qquad \text{Q.E.D.}$$

(3.1.2) PROPOSITION. *With* U, ζ_1, ζ, a, *and* k *as in Proposition* (3.1.1), *suppose* $\zeta_1 a \in H_s^{p,q}$ *for some* $s > 0$. *If* ϕ *satisfies* $(F-\lambda)\phi = a$ *for some constant* λ, *then* $\zeta\phi \in H_{s+k}^{p,q}$.

Proof: Assume $k = 1$; the proof in the other case is the same. Set $a' = a + \lambda\phi$, so that $F\phi = a'$; then since $a' \in H_0^{p,q}$ we have $\zeta_1\phi \in H_1^{p,q}$ by Proposition (3.1.1). Let $\{\zeta_j\}_2^s$ be a sequence of smooth functions with $\zeta_s = \zeta$ and $\zeta_j = 1$ on supp ζ_{j+1}. Then, using Proposition (3.1.1) and proceeding inductively,

$$\zeta_1 a' \in H_1^{p,q} \Longrightarrow \zeta_2\phi \in H_2^{p,q} \Longrightarrow \zeta_2 a' \in H_2^{p,q} \Longrightarrow \zeta_3\phi \in H_3^{p,q}, \text{ etc.} \qquad \text{Q.E.D.}$$

(3.1.3) COROLLARY. *If* $a|U \in \Lambda^{p,q}(U \cap \bar{M})$ *then* $\phi|U \in \Lambda^{p,q}(U \cap \bar{M})$. *In other words,* $F - \lambda$ *is hypoelliptic for every* λ.

Proof: Apply the Sobolev lemma (cf. Appendix, §2) to Proposition (3.1.2). Q.E.D.

(3.1.4) PROPOSITION. *If* $F\phi = a$ *and* $a \in \Lambda^{p,q}(\bar{M})$ *then* $\phi \in \Lambda^{p,q}(\bar{M})$ *and* $\|\phi\|_{s+1}^2 \lesssim \|a\|_s^2$ *for each* s.

Proof: This follows immediately from the Main Theorem by taking $U = \bar{M}$ and noting that $\|a\| \le \|a\|_s$. Q.E.D.

(3.1.5) COROLLARY. *If* $F\phi = a$ *and* $a \in H_s^{p,q}$ *then* $\phi \in H_{s+1}^{p,q}$ *and* $\|\phi\|_{s+1}^2 \lesssim \|a\|_s$.

Proof: Approximate a by smooth forms and use Proposition (3.1.4). Q.E.D.

(3.1.6) COROLLARY. F^{-1} *is a compact (completely continuous) operator.*

Proof: By Corollary (3.1.5), F^{-1} is bounded from $H_0^{p,q}$ to $H_1^{p,q}$, so the assertion follows from the Rellich lemma (cf. Appendix, §2). Q.E.D.

(3.1.7) COROLLARY. F *has a discrete spectrum with no finite limit point, and each eigenvalue occurs with finite multiplicity.*

Proof: Apply the theory of compact operators (cf. [39]) to Corollary (3.1.6). Q.E.D.

(3.1.8) COROLLARY. Q *is compact, that is, every* Q*-bounded sequence in* $\tilde{\mathfrak{D}}^{p,q}$ *has a subsequence converging in* $H_0^{p,q}$.

Proof: Since F is a positive self-adjoint operator, it has a positive square root $F^{1/2}$, and if $\phi \in \mathrm{Dom}\,(F)$, $Q(\phi,\phi) = (F\phi,\phi) = (F^{1/2}\phi, F^{1/2}\phi)$. Thus since $\mathrm{Dom}\,(F)$ is dense in $\tilde{\mathfrak{D}}^{p,q}$, $\tilde{\mathfrak{D}}^{p,q} \subset \mathrm{Dom}\,(F^{1/2})$ and $Q(\phi,\phi) = (F^{1/2}\phi, F^{1/2}\phi)$ for all $\phi \in \tilde{\mathfrak{D}}^{p,q}$. But $F^{-1/2} = (F^{1/2})^{-1}$ is compact (by Corollary (3.1.7), it is the norm limit of operators of finite rank), and this proves the assertion. (For an alternative proof, cf. the Appendix, §3.) Q.E.D.

(3.1.9) COROLLARY. *The eigenforms of* F *are all smooth.*

Proof: If ϕ is an eigenform with eigenvalue λ, then $(F-\lambda)\phi = 0$. Apply Corollary (3.1.3) with $U = \overline{M}$. Q.E.D.

At this point we pause to give the promised proof that $F = F_1$ (cf. §1.3).

(3.1.10) PROPOSITION. $F = F_1$.

Proof: First, it follows from Propositions (1.3.2) and (1.3.5) that $\mathrm{Dom}\,(F) \cap \Lambda^{p,q}(\overline{M}) = \mathrm{Dom}\,(F_1) \cap \Lambda^{p,q}(\overline{M}) = \{\phi \in \mathfrak{D}^{p,q} : \bar{\partial}\phi \in \mathfrak{D}^{p,q+1}\}$ and

$F = F_1$ on this domain. Suppose $\phi \in \mathrm{Dom}(F)$ and $F\phi = \alpha$. Choose a sequence $\{\alpha_n\} \subset \Lambda^{p,q}(\bar M)$ converging to α in $H_0^{p,q}$, and set $\phi_n = F^{-1}\alpha_n$; then $\phi_n \to \phi$ in $H_0^{p,q}$ and, by Proposition (3.1.4), $\phi_n \in \Lambda^{p,q}(\bar M)$. But then $F_1\phi_n = F\phi_n \to \alpha$, so since F_1 is a closed operator, $\phi \in \mathrm{Dom}(F_1)$ and $F_1\phi = \alpha$. Thus F_1 is an extension of F, and since F and F_1 are self-adjoint they must therefore be equal. Q.E.D.

It is high time we weaned ourselves away from the operator F and got back to the main business at hand, namely the study of the operators $\bar\partial$, ϑ, and $\square$. From the discussion of §1.3 we see that the operators $\square_F = F - I$ and $\bar\partial^*$ are the restrictions of $\square$ and ϑ (considered as acting on distributions) to $\mathrm{Dom}(F)$ and $\mathrm{Dom}(\bar\partial^*)$, respectively.

(3.1.11) PROPOSITION. $H_0^{p,q}$ *has a complete orthonormal basis of eigenforms for* $\square_F$ *which are smooth up to the boundary. The eigenvalues are non-negative, have no finite limit point, and occur with finite multiplicity. Moreover, for each* s, $\|\phi\|_{s+1}^2 \lesssim \|\square\phi\|_s^2 + \|\phi\|^2$ *uniformly for* $\phi \in \mathrm{Dom}(F) \cap \Lambda^{p,q}(\bar M)$. *(The corresponding localized estimates also hold, of course.)*

Proof: The first assertions follow from Corollaries (3.1.7) and (3.1.9). The estimates follow by induction on s from Proposition (3.1.4), since

$$\|\phi\|_1^2 \lesssim \|F\phi\|^2 \le \|\square\phi\|^2 + \|\phi\|^2 ,$$

$$\|\phi\|_2^2 \lesssim \|F\phi\|_1^2 \le \|\square\phi\|_1^2 + \|\phi\|_1^2 \le \|\square\phi\|_1^2 + \|\square\phi\|^2 + \|\phi\|^2$$

$$\lesssim \|\square\phi\|_1^2 + \|\phi\|^2 ,$$

and so forth. Q.E.D.

Remarks: (1) The s-norm estimate clearly extends to the case $\phi \in H_s^{p,q}$.

(2) In the case where M is the unit ball in C^n, Folland [9] has obtained the explicit eigenform decomposition of $H_0^{0,q}$ $(0 \le q \le n)$ and has shown by estimates on the eigenvalues that the estimate $\|\phi\|_1^2 \lesssim \|\square\phi\|^2 + \|\phi\|^2$ is sharp.

We shall denote the nullspace of an operator T by $\mathcal{N}(T)$.

By Proposition (3.1.11), the *harmonic space* $\mathcal{H}^{p,q} = \mathcal{N}(\square_F)$ is a finite-dimensional subspace of $\Lambda^{p,q}(\overline{M})$. Moreover, since $\square_F$ is bounded away from zero on the orthogonal complement $(\mathcal{H}^{p,q})^{\perp}$, Range $(\square_F)$ is closed. As $\square_F$ is self-adjoint, we obtain:

(3.1.12) PROPOSITION. $H_0^{p,q}$ *admits the strong orthogonal decomposition*

$$H_0^{p,q} = \text{Range}\,(\square_F) \oplus \mathcal{H}^{p,q} = \overline{\partial}\vartheta\,\text{Dom}\,(F) \oplus \vartheta\overline{\partial}\,\text{Dom}\,(F) \oplus \mathcal{H}^{p,q}\,.$$

Proof: The first equality follows from the fact that Range $(\square_F) = \mathcal{N}(\square_F)^{\perp}$, and the second from the fact that $\overline{\partial}^2 = 0$ and hence Range $(\overline{\partial}) \perp$ Range $(\overline{\partial}^*)$. Q.E.D.

(3.1.13) COROLLARY. *The range of* $\overline{\partial}$ *on* Dom $(\overline{\partial}) \cap H_0^{p,q-1}$ *is closed.*

Proof: Since Range $(\overline{\partial}) \perp \mathcal{N}(\overline{\partial}^*)$, we have Range $(\overline{\partial}) = \overline{\partial}\vartheta\,$Dom (F). Q.E.D.

We define the harmonic projector H to be the orthogonal projection of $H_0^{p,q}$ onto $\mathcal{H}^{p,q}$, and we define the *Neumann operator* $N{:}H_0^{p,q} \to \text{Dom}\,(F)$ as follows. If $a \in \mathcal{H}^{p,q}$, $Na = 0$; if $a \in \text{Range}\,(\square_F)$, $Na = \phi$ where ϕ is the unique solution of $\square_F\phi = a$ with $\phi \perp \mathcal{H}^{p,q}$, and we extend by linearity. In other words, Na is the unique solution ϕ of the equations $H\phi = 0$, $\square_F\phi = a - Ha$. (Note that N is defined whenever Range $(\square_F)$ is closed, whether the estimates hold or not.) We then have the following result, which is the *solution of the* $\overline{\partial}$*-Neumann problem* in the sense usually proposed in the literature ([22], [41]).

(3.1.14) THEOREM.

(1) N *is a compact operator.*

(2) *For any* $a \in H_0^{p,q}$, $a = \overline{\partial}\vartheta Na + \vartheta\overline{\partial}Na + Ha$.

(3) $NH = HN = 0$, $N\square = \square N = I - H$ *on* Dom (F), *and if* N *is also defined on* $H_0^{p,q+1}$ *(resp.* $H_0^{p,q-1}$*), then* $N\overline{\partial} = \overline{\partial}N$ *on* Dom $(\overline{\partial})$ *(resp.* $N\vartheta = \vartheta N$ *on* Dom $(\overline{\partial}^*)$*).*

(4) $N(\Lambda^{p,q}(\bar{M})) \subset \Lambda^{p,q}(\bar{M})$, *and for each* s *the estimate* $\|N\alpha\|_{s+1} \lesssim \|\alpha\|_s$ *holds for all* $\alpha \in \Lambda^{p,q}(\bar{M})$.

Proof: (1) follows from Proposition (3.1.11). (2) is a restatement of Proposition (3.1.12). As for (3), it is immediate from the definition of N that $NH = HN = 0$ and $N\square = \square N = I - H$ on Dom (F). Moreover, if $\alpha \in \text{Dom}(\bar{\partial})$ and N is also defined on $H_0^{p,q+1}$, we have by (2) and the first assertions of (3), $N\bar{\partial}\alpha = N\bar{\partial}\vartheta\bar{\partial}N\alpha = N\square\bar{\partial}N\alpha = (I-H)\bar{\partial}N\alpha = \bar{\partial}N\alpha$ since $\bar{\partial}^2 = 0$ and $H\bar{\partial} = 0$. Likewise for ϑ. Finally, the first assertion of (4) follows from Corollary (3.1.8) since $\alpha - H\alpha$ is smooth whenever α is, and the second follows from Proposition (3.1.11) and (3), as

$$\|N\alpha\|_{s+1}^2 \lesssim \|\square N\alpha\|_s^2 + \|N\alpha\|^2 \lesssim \|\alpha\|_s^2 + \|H\alpha\|_s^2 + \|N\alpha\|^2$$

$$\lesssim \|\alpha\|_s^2 + \|H\alpha\|^2 + \|N\alpha\|^2 \lesssim \|\alpha\|_s^2$$

(all s-norms on the finite-dimensional space $\mathcal{H}^{p,q}$ being equivalent). Q.E.D.

Remark. The estimates of (4) are clearly localizable.

We now turn to the problem of solving the inhomogeneous Cauchy-Riemann equation $\bar{\partial}\phi = \alpha$. Note first of all that there is no hope of solving this equation unless $\alpha \perp \mathcal{N}(\bar{\partial}^*)$, or equivalently, $\bar{\partial}\alpha = 0$ and $H\alpha = 0$.

(3.1.15) PROPOSITION. *Suppose* $\alpha \in H_0^{p,q}$, $\bar{\partial}\alpha = 0$, *and* $H\alpha = 0$. *Then there is a unique solution* ϕ *of* $\bar{\partial}\phi = \alpha$ *with* $\phi \perp \mathcal{N}(\bar{\partial})$. *If* $\alpha \in \Lambda^{p,q}(\bar{M})$ *then* $\phi \in \Lambda^{p,q-1}(\bar{M})$ *and* $\|\phi\|_s \lesssim \|\alpha\|_s$ *for each* s.

Proof: By Theorem (3.1.14) and the conditions on α, we have $\alpha = \bar{\partial}\vartheta N\alpha$. Thus we may take $\phi = \vartheta N\alpha$, and the condition $\phi \perp \mathcal{N}(\bar{\partial})$ clearly implies uniqueness. Moreover, if $\alpha \in \Lambda^{p,q}(\bar{M})$ then so is $N\alpha$, and hence $\phi \in \Lambda^{p,q-1}(\bar{M})$; and we have $\|\phi\|_s \lesssim \|N\alpha\|_{s+1} \lesssim \|\alpha\|_s$. Q.E.D.

We can actually obtain sharper estimates for the solution $\phi = \vartheta N\alpha$.

(3.1.16) PROPOSITION. *Under the conditions of Proposition* (3.1.15), *let* U *be a subregion of* $\bar{M}$ *and* $\zeta \in \Lambda_0^{0,0}(U)$. *If* $U \cap bM = \emptyset$, $\|\zeta\phi\|_{s+1} \lesssim \|a\|_s$, *and if* U *is a special boundary chart,* $\Sigma_{|\beta|=s+1} |||D^\beta \zeta\phi|||_{-1/2} \lesssim \|a\|_s$.

Proof: In the first case we have

$$\|\zeta\phi\|_{s+1} \le \|\vartheta\zeta N a\|_{s+1} + \|\zeta' N a\|_{s+1} \lesssim \|\zeta N a\|_{s+2} + \|\zeta' N a\|_{s+1} \;;$$

it suffices to estimate $\|\zeta N a\|_{s+2}$, as the same estimate holds for $\|\zeta' N a\|_{s+1}$. But since $Ha = 0$, $\Box N a = a$, and we have by Theorem 2.2.5,

$$\|\zeta N a\|_{s+2} \lesssim \|F N a\|_s = \|(\Box + I) N a\|_s \le \|a\|_s + \|N a\|_s \lesssim \|a\|_s \,.$$

For the boundary estimate, we only sketch the proof. Assume first that $s = 0$. For any tangential derivative we have $|||D_t^j \zeta\phi|||_{-1/2}^2 \lesssim |||\zeta\phi|||_{1/2}^2 = (\Lambda_t^{1/2} \zeta\vartheta N a, \Lambda_t^{1/2} \zeta\vartheta N a)$. Switching around the operators in this expression, we see by the usual estimates that

$$\begin{aligned}
(\Lambda_t^{1/2} \zeta\vartheta N a, \Lambda_t^{1/2} \zeta\vartheta N a) &= (\zeta\bar{\partial}\vartheta N a, \Lambda_t^1 \zeta N a) + \mathcal{O}(\|a\|^2) \\
&= (\zeta a, \Lambda_t^1 \zeta N a) + \mathcal{O}(\|a\|^2) \\
&\lesssim \|a\| \; |||\zeta N a|||_1 + \mathcal{O}(\|a\|^2) \\
&\lesssim \|a\| \; \|N a\|_1 + \mathcal{O}(\|a\|^2) \\
&\lesssim \|a\|^2,
\end{aligned}$$

To obtain the estimate for $D_r\phi$, we note that the system

$$\bar{\partial} \oplus \vartheta : \Lambda^{p,q}(\bar{M}) \to \Lambda^{p,q+1}(\bar{M}) \oplus \Lambda^{p,q-1}(\bar{M})$$

is elliptic; hence by the inversion argument used in Theorems (2.3.4) and (2.4.8), $D_r\phi$ can be expressed in terms of $\bar{\partial}\phi$, $\vartheta\phi$, and $D_t^j\phi$. But $\vartheta\phi = \vartheta^2 N a = 0$, so $D_r\phi$ is expressed in terms of $\bar{\partial}\phi = a$ and $D_t^j\phi$, which yields the estimate for $D_r\phi$. The theorem for general s is now proved by the usual sort of inductive argument; details are left to the reader. Q.E.D.

If the basic estimate fails in $\mathcal{D}^{p,q}$ but holds in $\mathcal{D}^{p,q-1}$ and $\mathcal{D}^{p,q+1}$, we can still obtain a lot of information about (p,q)-forms. First, we can still speak of the harmonic space $\mathcal{H}^{p,q} = \mathcal{N}(\Box_F)$; since $\Box_F$ is a closed operator, $\mathcal{H}^{p,q}$ is closed and hence the projection H also makes sense.

(3.1.17) **Proposition**. *Suppose the basic estimate holds in* $\mathcal{D}^{p,q-1}$ *and* $\mathcal{D}^{p,q+1}$ *but not necessarily in* $\mathcal{D}^{p,q}$. *If* $\phi \in \tilde{\mathcal{D}}^{p,q}$ *then* $H\phi = \phi - \vartheta N\bar\partial\phi - \bar\partial N\vartheta\phi$.

Proof: Let $K\phi = \phi - \vartheta N\bar\partial\phi - \bar\partial N\vartheta\phi$; to prove that $K = H$ it suffices to show $K\phi \in \mathcal{H}^{p,q}$ and $\phi - K\phi \perp \mathcal{H}^{p,q}$. Now

$$\bar\partial K\phi = \bar\partial\phi - \bar\partial\vartheta N\bar\partial\phi = \bar\partial\phi - \Box_F N\bar\partial\phi = \bar\partial\phi - \bar\partial\phi = 0$$

since $\bar\partial\phi \in \mathcal{N}(\bar\partial) \cap (\mathcal{H}^{p,q-1})^\perp$. Likewise $\vartheta K\phi = 0$; here we need $\phi \in \tilde{\mathcal{D}}^{p,q}$ to guarantee $\vartheta\phi = \bar\partial^*\phi \in \mathcal{N}(\bar\partial^*) \cap (\mathcal{H}^{p,q-1})^\perp$. $\phi \in \tilde{\mathcal{D}}^{p,q}$ also implies $\vartheta K\phi = \bar\partial^* K\phi$, so $K\phi \in \mathcal{N}(\bar\partial) \cap \mathcal{N}(\bar\partial^*) = \mathcal{H}^{p,q}$. On the other hand, $\phi - K\phi = \vartheta N\bar\partial\phi + \bar\partial N\vartheta\phi \in \text{Range}\,(\bar\partial^*) \oplus \text{Range}\,(\bar\partial) = (\mathcal{H}^{p,q})^\perp$. Q.E.D.

(3.1.18) **Proposition**. *Suppose the basic estimate holds in* $\mathcal{D}^{p,q-1}$ *and* $\mathcal{D}^{p,q+1}$ *but not necessarily in* $\mathcal{D}^{p,q}$. *Then* $\Box_F$ *is bounded away from zero on* $(\mathcal{H}^{p,q})^\perp$.

Proof: Suppose $\phi \in \text{Dom}\,(\Box_F) \cap (\mathcal{H}^{p,q})^\perp$. Write $\phi = \phi_1 + \phi_2$ where $\phi_1 \perp \mathcal{N}(\bar\partial)$ and $\phi_2 \perp \mathcal{N}(\bar\partial^*)$. By Proposition (3.1.15) we have $\phi_1 = \vartheta N\bar\partial\phi$ and hence $\|\phi_1\|^2 \lesssim \|\bar\partial\phi\|^2$. The reasoning of Proposition (3.1.15) applied to $\bar\partial^*$ also shows that $\phi_2 = \bar\partial N\vartheta\phi$ and hence $\|\phi_2\|^2 \lesssim \|\vartheta\phi\|^2$. Now $Q(\psi,\psi) - (\psi,\psi)$ is a positive definite form on $(\mathcal{H}^{p,q+1})^\perp$ by Proposition (3.1.11) since $Q(\psi,\psi) - (\psi,\psi) = (\Box_F\psi,\psi) \gtrsim \|\psi\|^2$; thus

$$\|\bar\partial\phi\|^2 \lesssim Q(\bar\partial\phi,\bar\partial\phi) - (\bar\partial\phi,\bar\partial\phi) = \|\vartheta\bar\partial\phi\|^2 .$$

Likewise, $\|\vartheta\phi\|^2 \lesssim \|\bar\partial\vartheta\phi\|^2$. Adding these estimates, we obtain $\|\phi\|^2 = \|\phi_1\|^2 + \|\phi_2\|^2 \lesssim \|\vartheta\bar\partial\phi\|^2 + \|\bar\partial\vartheta\phi\|^2 = \|\Box_F\phi\|^2$. Q.E.D.

As a result of Proposition (3.1.18), Range ($\square_F$) is closed, so we obtain the strong orthogonal decomposition of $H_0^{p,q}$. We can also define the Neumann operator N just as before, and we have the following analogue of Theorem (3.1.14):

(3.1.19) THEOREM. *Suppose the basic estimate holds in* $\mathfrak{D}^{p,q-1}$ *and* $\mathfrak{D}^{p,q+1}$ *but not necessarily in* $\mathfrak{D}^{p,q}$. *Then*:

(1) N *is bounded.*

(2) *For any* $\alpha \in H_0^{p,q}$, $\alpha = \bar{\partial}\vartheta N\alpha + \vartheta\bar{\partial}N\alpha + H\alpha$.

(3) $NH = HN = 0$, $N\square = \square N = I - H$ *on* Dom (F), *and* N *commutes with* $\bar{\partial}$ *and* ϑ *on* Dom ($\bar{\partial}$) *and* Dom ($\bar{\partial}^*$), *respectively.*

(4) $H(\mathfrak{D}^{p,q}) \subset \mathfrak{D}^{p,q}$ *and* $N(\mathfrak{D}^{p,q}) \subset \mathfrak{D}^{p,q}$.

Proof: (1) and (2) are immediate from Proposition (3.1.18), and (3) follows just as in Theorem (3.1.14). The fact that $H(\mathfrak{D}^{p,q}) \subset \mathfrak{D}^{p,q}$ follows from the formula of Proposition (3.1.17) and the regularity of N at levels (p,q−1) and (p,q+1). Finally, if $\alpha \in \mathfrak{D}^{p,q} \subset$ Dom ($\bar{\partial}^*$) we have, by (2) and (3) and Theorem (3.1.14),

$$N\alpha = N\bar{\partial}\vartheta N\alpha + N\vartheta\bar{\partial}N\alpha = N\bar{\partial}N\vartheta\alpha + N\vartheta N\bar{\partial}\alpha$$

$$= \bar{\partial}N^2\vartheta\alpha + \vartheta N^2\bar{\partial}\alpha \in \mathfrak{D}^{p,q}$$

– again, by virtue of the regularity of N at levels (p,q−1) and (p,q+1). Q.E.D.

Remarks: (1) Conclusion (4) becomes particularly cogent in case $q = 0$, for $\mathfrak{D}^{p,0} = \Lambda^{p,0}(\bar{M})$. In any event, note that Proposition (3.1.18) and Theorem (3.1.19) are remarkably strong regularity results for $\square_F$ on (p,q)-forms considering that we have no estimates for this operator!

(2) It is well-known that the harmonic projection H on functions can be represented as $Hu(x) = \int_M K(x,y)\overline{u(y)}\,dy$ where $K(x,y) = \overline{K(y,x)}$ is the Bergman kernel function, which is smooth on $M \times M$. Suppose the basic estimate holds in $\mathfrak{D}^{0,1}$. By Proposition (3.1.17), H has the

property that Hu is smooth wherever u is. In particular, H maps $\Lambda^{0,0}(\bar{M})$ into itself, and this mapping is continuous in the $\mathcal{C}^\infty$ topology. Since H is self-adjoint, it extends to a continuous mapping of the dual space $\mathcal{D}'$ of $\Lambda^{0,0}(\bar{M})$ (= the space of distributions on $\bar{M}$) into itself, and this extension clearly preserves the local smoothness property. In particular, $K(x,y) = (H\delta_y)(x)$ where $\delta_y \in \mathcal{D}'$ is the Dirac "function" defined by $\langle\delta_y, u\rangle = u(y)$. It then follows that $K(x,y)$ is smooth on $\bar{M} \times \bar{M}$ except along the boundary diagonal $\Delta = \{(x,x) : x \in bM\}$. This result is due to N. Kerzman [21].

2. *Pseudoconvexity and the basic estimate*

We would be poorly rewarded for having traveled this long journey if we had no way of knowing when the basic estimate holds. In §2.1 we hinted that strong pseudoconvexity is a sufficient condition for the basic estimate to hold in $\mathcal{D}^{p,q}$, and indeed this is the case for $q \neq 0$. But the remarkable fact, which was discovered by L. Hörmander [16], is that the basic estimate in $\mathcal{D}^{p,q}$ is completely equivalent to a geometric condition on bM related to pseudoconvexity. In this section we present Hörmander's arguments.

For each $p \in bM$, the *Levi form* at p is the Hermitian form on the (n−1)-dimensional space $(\Pi_{1,0}CT_pM) \cap CT_pbM$ given by

$$(L_1, L_2) \leftrightarrow 2\langle\partial\bar\partial r, L_1 \wedge \bar{L}_2\rangle .$$

(It is Hermitian because $\partial\bar\partial = -\bar\partial\partial = -\overline{\partial\bar\partial}$.) The following proposition shows that this form depends only on the outward normal dr to bM and is therefore intrinsically defined.

We shall be working in special boundary charts U, with the special basis $\{\omega_i\}_1^n$, $\omega_n = \sqrt{2}\,\partial r$, for $\Lambda^{1,0}(U)$. If $L_1, \ldots, L_n$ are the dual vector fields, then $\{(L_i)_p\}_1^{n-1}$ is an orthonormal basis for the space $(\Pi_{1,0}CT_pM) \cap CT_pbM$ on which the Levi form is defined. With respect to this basis, we define the matrix coefficients of the Levi form, $c_{ij} = 2\langle\partial\bar\partial r, L_i \wedge \bar{L}_j\rangle$.

(3.2.1) PROPOSITION. $c_{ij} = \frac{1}{\sqrt{2}} <\omega_n, [L_i, \bar{L}_j]>$.

Proof: By the formula for $d\xi$ where ξ is a one-form,

$$\begin{aligned} c_{ij} &= 2<\partial\bar{\partial}r, L_i \wedge \bar{L}_j> = 2<d\bar{\partial}r, L_i \wedge \bar{L}_j> \\ &= L_i(<\bar{\partial}r, L_j>) - L_j(<\bar{\partial}r, L_i>) - <\bar{\partial}r, [L_i, \bar{L}_j]>. \end{aligned}$$

But $<\bar{\partial}r, L_i> = 0$, and $<\bar{\partial}r, \bar{L}_j> = 0$ since $j \leq n-1$, so this expression reduces to $-<\bar{\partial}r, [L_i, \bar{L}_j]>$. Now $L_i r = \bar{L}_j r = 0$, so $<dr, [L_i, \bar{L}_j]> = [L_i, \bar{L}_j]r = 0$, hence $<\bar{\partial}r, [L_i, \bar{L}_j]> = -<\partial r, [L_i, \bar{L}_j]>$, and we finally obtain $c_{ij} = <\partial r, [L_i, \bar{L}_j]> = \frac{1}{\sqrt{2}}<\omega_n, [L_i, \bar{L}_j]>$. Q.E.D.

In other words, c_{ij} is the coefficient of L_n in the basis expansion of $\frac{1}{\sqrt{2}}[L_i, \bar{L}_j]$, and it is this property which is crucial for our purposes. Note that by a proper choice of $\omega_1, \ldots, \omega_{n-1}$, we can always arrange that the Levi form be diagonal at a given point p, i.e., $c_{ij} = \lambda_i \delta_{ij} + b_{ij}$ where $b_{ij}(p) = 0$.

M is said to be *pseudoconvex* if the Levi form is positive semi-definite at each point of bM and *strongly pseudoconvex* if it is positive definite at each point of bM. We say that M satisfies *condition* Z(q) if the Levi form has at least $n-q$ positive eigenvalues or at least $q+1$ negative eigenvalues at each point of bM. (Thus a strongly pseudoconvex manifold satisfies condition Z(q) for all $q > 0$.)

(3.2.2) THEOREM. *The basic estimate holds in* $\mathfrak{D}^{p,q}$ *if and only if* M *satisfies condition* Z(q).

We shall prove the forward and reverse implications of Theorem (3.2.2) separately, using a series of lemmas.

(3.2.3) LEMMA. *Given* $p \in bM$, *let* U *be a special boundary chart near* p *such that the Levi form at* p *is diagonal with eigenvalues* $\lambda_1, \ldots, \lambda_{n-1}$. *Then for any* $\delta > 0$ *there is a neighborhood* $V \subset U$ *of* p *such that*

$$Q(\phi,\phi) = \sum_{kIJ} \|\bar{L}_k\phi_{IJ}\|^2 + \sum_{IJ}\sum_{k\epsilon J} \lambda_k \int_{bM} |\phi_{IJ}|^2 + R(\phi) + \mathcal{O}(E(\phi)\|\phi\|)$$

uniformly for $\phi \in \mathcal{D}^{p,q} \cap \Lambda_0^{p,q}(V \cap \bar{M})$, *where* $|R(\phi)| \le \delta E(\phi)^2$. *(Note that the second term on the right makes sense because if* $n \in J$, $\phi_{IJ} = 0$ *on* bM *by Lemma* (2.3.2).)

Proof: From formula (1.2.3) we have

$$\|\bar{\partial}\phi\|^2 = \sum_{IJ}\sum_{k\notin J} \|\bar{L}_k\phi_{IJ}\|^2 + \sum_{\substack{k\neq m\\ \langle kJ\rangle=\langle mS\rangle}} \sum_I \epsilon^{\langle kJ\rangle}_{kJ}\,\epsilon^{\langle mS\rangle}_{mS}\,(\bar{L}_k\phi_{IJ},\bar{L}_m\phi_{IS})$$
$$+ \mathcal{O}\Big(\sum_{kIJ} \|\bar{L}_k\phi_{IJ}\|\,\|\phi\| + \|\phi\|^2\Big)$$
$$= \sum_{kIJ} \|\bar{L}_k\phi_{IJ}\|^2 - \sum_{IJ}\sum_{k\epsilon J} \|\bar{L}_k\phi_{IJ}\|^2$$
$$+ \sum_{\substack{k\neq m\\ \langle kJ\rangle=\langle mS\rangle}} \sum_I \epsilon^{\langle kJ\rangle}_{kJ}\,\epsilon^{\langle mS\rangle}_{mS}\,(\bar{L}_k\phi_{IJ},\bar{L}_m\phi_{IS}) + \mathcal{O}(E(\phi)\|\phi\|)$$

where, for example, $\langle kJ\rangle$ is the strictly increasing (q+1)-tuple whose elements are $k, j_1, \ldots, j_q$. Now if $\langle kJ\rangle = \langle mS\rangle$ we have $J = \langle mH\rangle$ and $S = \langle kH\rangle$ for some H, and it is easily verified that $\epsilon^{\langle kJ\rangle}_{kJ}\,\epsilon^{\langle mS\rangle}_{mS} = -\epsilon^{J}_{mH}\,\epsilon^{S}_{kH}$, so the third term on the right equals $-\Sigma_{IH}\,\Sigma_{k\neq m}\,\epsilon^{J}_{mH}\,\epsilon^{S}_{kH}\,(\bar{L}_k\phi_{IJ}, \bar{L}_m\phi_{IS})$. The corresponding sum for $m = k$ is just the second term on the right, and hence

$$\|\bar{\partial}\phi\|^2 = \sum_{kIJ} \|\bar{L}_k\phi_{IJ}\|^2 - \sum_{mkIH} \epsilon^{J}_{mH}\,\epsilon^{S}_{kH}\,(\bar{L}_k\phi_{IJ}, \bar{L}_m\phi_{IS}) + \mathcal{O}(E(\phi)\|\phi\|). \tag{3.2.4}$$

We proceed to manipulate the cross terms. First,

$$-(\bar{L}_k\phi_{IJ}, \bar{L}_m\phi_{IS}) = (L_m\bar{L}_k\phi_{IJ}, \phi_{IS}) + \mathcal{O}(E(\phi)\|\phi\|)\,;$$

there is no boundary term since $\bar{L}_m$ is tangential when $m < n$ and $\phi_{IJ} = 0$ on bM when $m = n$ by Lemma (2.3.2). Next,

$$(L_m \bar{L}_k \phi_{IJ}, \phi_{IS}) = (\bar{L}_k L_m \phi_{IJ}, \phi_{IS}) + ([L_m, \bar{L}_k]\phi_{IJ}, \phi_{IS})$$
$$= -(L_m \phi_{IJ}, L_k \phi_{IS}) + ([L_m, \bar{L}_k]\phi_{IJ}, \phi_{IS}) + \mathcal{O}(E(\phi)\|\phi\|),$$

the boundary terms vanishing as before. Therefore,

$$-\sum_{mkIH} \epsilon^J_{mH} \epsilon^S_{kH} (\bar{L}_k \phi_{IJ}, \bar{L}_m \phi_{IS})$$
$$= -\sum_{mkIH} \epsilon^J_{mH} \epsilon^S_{kH} (L_m \phi_{IJ}, L_k \phi_{IS})$$
$$+ \sum_{mkIH} \epsilon^J_{mH} \epsilon^S_{kH} ([L_m, \bar{L}_k]\phi_{IJ}, \phi_{IS}) + \mathcal{O}(E(\phi)\|\phi\|).$$

But from formula (1.2.4), the first term on the right is $-\|\vartheta\phi\|^2 + \mathcal{O}(E(\phi)\|\phi\|)$. Substituting in (3.2.4), therefore,

$$\|\bar{\partial}\phi\|^2 = \sum_{kIJ} \|\bar{L}_k \phi_{IJ}\|^2 - \|\vartheta\phi\|^2 + \mathcal{O}(E(\phi)\|\phi\|)$$
$$+ \sum_{mkIH} \epsilon^J_{mH} \epsilon^S_{kH} ([L_m, \bar{L}_k]\phi_{IJ}, \phi_{IS}),$$

or

$$Q(\phi, \phi) = \sum_{kIJ} \|\bar{L}_k \phi_{IJ}\|^2 + \mathcal{O}(E(\phi)\|\phi\|) \tag{3.2.5}$$
$$+ \sum_{mkIH} \epsilon^J_{mH} \epsilon^S_{kH} ([L_m, \bar{L}_k]\phi_{IJ}, \phi_{IS}).$$

It remains therefore to estimate the last term on the right of (3.2.5). Let $[L_m, \bar{L}_k] = \Sigma_j (a^j_{mk} L_j + b^j_{mk} \bar{L}_j)$. If $j < n$ we have

$$(a^j_{mk} L_j \phi_{IJ}, \phi_{IS}) + (b^j_{mk} \bar{L}_j \phi_{IJ}, \phi_{IS})$$
$$= -(a^j_{mk} \phi_{IJ}, \bar{L}_j \phi_{IS}) + (b_{mk} \bar{L}_j \phi_{IJ}, \phi_{IS}) + \mathcal{O}(E(\phi)\|\phi\|)$$
$$= \mathcal{O}(E(\phi)\|\phi\|).$$

Now if $m = n$ or $k = n$ we have $\phi_{IJ} = 0$ on bM or $\phi_{IS} = 0$ on bM, so

$$(a^n_{nk} L_n \phi_{IJ}, \phi_{IS}) = -(a^n_{nk}\phi_{IJ}, \bar{L}_n \phi_{IS}) + \mathcal{O}(E(\phi)\|\phi\|) = \mathcal{O}(E(\phi)\|\phi\|)$$

and

$$(a^n_{mn} L_n \phi_{IJ}, \phi_{IS}) = -(a^n_{mn}\phi_{IJ}, \bar{L}_n \phi_{IS}) + \mathcal{O}(E(\phi)\|\phi\|) = \mathcal{O}(E(\phi)\|\phi\|)\,.$$

On the other hand, if $m, k \leq n-1$, $a^n_{mk} = \sqrt{2}\, c_{mk}$ by Proposition (3.2.1). Hence:

$$\text{(3.2.6)}\begin{cases} ([L_m, L_k]\phi_{IJ}, \phi_{IS}) = \mathcal{O}(E(\phi)\|\phi\|) \ \text{ if } \ m=n \ \text{ or } \ k=n, \ \text{ and} \\ ([L_m, \bar{L}_k]\phi_{IJ}, \phi_{IS}) = \sqrt{2}\,(c_{mk} L_n \phi_{IJ}, \phi_{IS}) + \mathcal{O}(E(\phi)\|\phi\|) \ \text{ if } \ m,k \leq n-1. \end{cases}$$

Next we write $c_{mk} = \lambda_k \delta_{mk} + b_{mk}$ where $b_{mk}(p) = 0$; then since $J = S$ when $m = k$,

$$\begin{aligned}\text{(3.2.7)}\quad (\lambda_k \delta_{mk} L_n \phi_{IJ}, \phi_{IS}) &= \lambda_k \delta_{mk} (L_n \phi_{IJ}, \phi_{IJ}) \\ &= \lambda_k \delta_{mk} (\phi_{IJ}, \bar{L}_n \phi_{IJ}) + \mathcal{O}(\|\phi\|^2) + \lambda_k \delta_{mk} \int_{bM} \sigma(L_n, dr) |\phi_{IJ} \\ &= \mathcal{O}(E(\phi)\|\phi\|) + \frac{1}{\sqrt{2}} \lambda_k \delta_{mk} \int_{bM} |\phi_{IJ}|^2\,,\end{aligned}$$

and

$$\begin{aligned}\text{(3.2.8)}\quad (b_{mk} L_n \phi_{IJ}, \phi_{IS}) &= (b_{mk} \phi_{IJ}, \bar{L}_n \phi_{IS}) + \mathcal{O}(\|\phi\|^2) + \int_{bM} \sigma(L_n, dr) b_{mk} \phi_{IJ} \overline{\phi_{IS}} \\ &= \mathcal{O}(E(\phi)\|\phi\|) + \frac{1}{\sqrt{2}} \int_{bM} b_{mk} \phi_{IJ} \overline{\phi_{IS}}\,.\end{aligned}$$

But

$$\left| \int_{bM} b_{mk} \phi_{IJ} \phi_{IS} \right| \leq \sup |b_{mk}| \int_{bM} |\phi|^2 \leq \sup |b_{mk}| E(\phi)^2$$

where the sup is taken over $\operatorname{supp} \phi$. Since $b_{mk}(p) = 0$, $\sup |b_{mk}|$ can be made arbitrarily small by taking $\operatorname{supp} \phi$ small.

Combining (3.2.6), (3.2.7), and (3.2.8) and summing, we therefore obtain

$$\sum_{mkIH} \epsilon^{J}_{mH}\, \epsilon^{S}_{kH}\, ([L_m,\bar{L}_k]\phi_{IJ},\phi_{IS}) = \sum_{IJ} \sum_{k\epsilon J} \lambda_k \int_{bM} |\phi_{IJ}|^2$$
$$+ R(\phi) + \mathcal{O}(E(\phi)\|\phi\|)$$

where $|R(\phi)| \leq \delta E(\phi)^2$ provided $\operatorname{supp}\phi$ is small. Substituting this in (3.2.5), we are done. Q.E.D.

(3.2.9) LEMMA. *Let* p, U, δ *be as in Lemma* (3.2.3). *There is a neighborhood* $V \subset U$ *of* p *such that for* $1 \leq k \leq n-1$,

$$\|\bar{L}_k u\|^2 \geq -\lambda_k \int_{bM} |u|^2 + \mathcal{O}(E(u)\|u\|) + R(u)$$

uniformly for $u \,\epsilon\, \Lambda_0^{0,0}(V \cap \bar{M})$, *where* $|R(u)| \leq \delta E(u)^2$.

Proof: By the same reasoning as in Lemma (3.2.3), we have

$$\begin{aligned}\|\bar{L}_k u\|^2 = (\bar{L}_k u, \bar{L}_k u) &= -(L_k\bar{L}_k u, u) + \mathcal{O}(E(u)\|u\|)\\ &= -([L_k, \bar{L}_k]u, u) + (L_k u, L_k u) + \mathcal{O}(E(u)\|u\|)\\ &\geq -([L_k, \bar{L}_k]u, u) + \mathcal{O}(E(u)\|u\|)\\ &= -\sqrt{2}(c_{kk}L_n u, u) + \mathcal{O}(E(u)\|u\|)\\ &= -\lambda_k \int_{bM} |u|^2 + \mathcal{O}(E(u)\|u\|) + R(u). \qquad \text{Q.E.D.}\end{aligned}$$

(3.2.10) THEOREM. *If* M *satisfies condition* $Z(q)$ *then the basic estimate holds in* $\mathfrak{D}^{p,q}$.

Proof: First let V be a special boundary chart on which the conclusions of Lemmas (3.2.3) and (3.2.9) hold, and consider $\phi \,\epsilon\, \mathfrak{D}^{p,q} \cap \Lambda_0^{p,q}(V \cap \bar{M})$. For any ϵ, $0 < \epsilon < 1$, by combining Lemmas (3.2.3) and (3.2.9) we see that

$$(3.2.11)\quad Q(\phi,\phi) = \sum_{kIJ} \|\bar{L}_k\phi_{IJ}\|^2 + \sum_{IJ}\sum_{k\epsilon J} \lambda_k \int_{bM} |\phi_{IJ}|^2 + R(\phi) + \mathcal{O}(E(\phi)\|\phi\|)$$

$$\geq \epsilon \sum_{kIJ} \|\bar{L}_k\phi_{IJ}\|^2 + (1-\epsilon)\sum_{IJ}\sum_{\lambda_k<0} \|\bar{L}_k\phi_{IJ}\|^2$$

$$+ \sum_{IJ}\sum_{k\epsilon J} \lambda_k \int_{bM} |\phi_{IJ}|^2 - \delta E(\phi)^2 - \mathcal{O}(E(\phi)\|\phi\|)$$

$$\geq \epsilon \sum_{kIJ} \|\bar{L}_k\phi_{IJ}\|^2 + (1-\epsilon)\sum_{IJ}\sum_{\lambda_k<0} (-\lambda_k) \int_{bM} |\phi_{IJ}|^2$$

$$+ \sum_{IJ}\sum_{k\epsilon J} \lambda_k \int_{bM} |\phi_{IJ}|^2 - 2\delta E(\phi)^2 - \mathcal{O}(E(\phi)\|\phi\|)\,.$$

Consider the situation for each fixed J. If $n \epsilon J$ then $\int_{bM} |\phi_{IJ}| = 0$; suppose then that $n \not\epsilon J$. If $n-q$ of the λ_k's are positive, there must be at least one positive λ_k, say λ_{k_0}, with $k_0 \epsilon J$. We can then choose ϵ small enough so that $\lambda_{k_0} \geq \epsilon\,(1 + \Sigma_{\lambda_k<0,k\epsilon J}(-\lambda_k))$. On the other hand, if $q+1$ of the λ_k's are negative, there must be at least one negative λ_k, say λ_{k_1}, with $k_1 \not\epsilon J$. We can then choose ϵ small enough so that $-\lambda_{k_1} \geq \frac{\epsilon}{1-\epsilon}(1 + \Sigma_{\lambda_k<0,k\epsilon J}(-\lambda_k))$. Taking ϵ so small that one of these estimates holds for each J and substituting into (3.2.11), we obtain

$$Q(\phi,\phi) \geq \epsilon\left(\sum_{kIJ} \|\bar{L}_k\phi_{IJ}\|^2 + \sum_{IJ} \int_{bM} |\phi_{IJ}|^2\right) - 2\delta E(\phi)^2 - \mathcal{O}(E(\phi)\|\phi\|)$$

$$\geq (\epsilon - 2\delta)\,E(\phi)^2 - \epsilon\|\phi\|^2 - \mathcal{O}(E(\phi)\|\phi\|)\,.$$

But $\mathcal{O}(E(\phi)\|\phi\|) \leq \delta E(\phi)^2 + (\ell c)\|\phi\|^2$, so taking $\delta = \frac{\epsilon}{4}$,

$$Q(\phi,\phi) \geq \frac{\epsilon}{4} E(\phi)^2 - (\ell c)\|\phi\|^2\,.$$

Since $\|\phi\|^2 \leq Q(\phi,\phi)$, this implies that $Q(\phi,\phi) \gtrsim E(\phi)^2$ uniformly for $\phi \,\epsilon\, \mathcal{D}^{p,q} \cap \Lambda^{p,q}(V \cap \bar{M})$.

Now cover bM by finitely many charts $\{V_j\}_1^N$ such that this conclusion holds on each chart, and choose V_0 so that $M - \cup_1^N V_j \subset V_0 \subset \bar{V}_0 \subset M$.

By Proposition (2.1.6) and Theorem (2.2.1), $Q(\phi,\phi) \gtrsim \|\phi\|_1^2 \gtrsim E(\phi)^2$ for all $\phi \in \Lambda_0^{p,q}(V_0)$. Using a partition of unity subordinate to $\{V_j\}_0^N$ together with Lemma (2.2.6), we are done. Q.E.D.

(3.2.12) COROLLARY. *If* M *is strongly pseudoconvex, the basic estimate holds in* $\mathfrak{D}^{p,q}$ *for* $q > 0$. *Hence the results of Theorem* (3.1.14) *apply for* $q > 0$ *and the results of Theorem* (3.1.19) *apply for* $q = 0$.

(3.2.13) COROLLARY. *The basic estimate always holds in* $\mathfrak{D}^{p,n}$.

For, condition Z(n) is vacuous. In fact, we can say more. The condition $\phi \in \mathfrak{D}^{p,n}$ is equivalent to the condition $\phi|bM = 0$; hence the $\bar{\partial}$-Neumann conditions for (p,n) forms reduce to the Dirichlet conditions, which are well known to be coercive, cf. [7]. Indeed, $\Lambda_0^{p,n}(M)$ is dense in $\mathfrak{D}^{p,n}$, so Gårding's inequality holds up to the boundary. The interior estimates of §2.2 are then valid at the boundary, the arguments of §2.3 apply with $\delta = 0$, and the delicate machinations of §2.4 are unnecessary. The coerciveness for $q = n$, as well as the non-coerciveness for $q < n$, is also a special case of a recent theorem of Sweeney [44] on overdetermined systems.

At this point we are in a position to prove the Newlander-Nirenberg theorem, and we shall do so in the following chapter. We shall assume this result in proving the converse part of Theorem (3.2.2), so that we can introduce complex analytic coordinate systems. First we prove a lemma in C^N.

(3.2.14) LEMMA. *Let* $L = \Sigma_1^N \lambda_j |z_j|^2$ *where* $\lambda_j \in \mathbf{R}$, *and set* $\lambda_j^- = \max(-\lambda_j, 0)$. *Then the inequality*

$$2c \int_{C^N} |\chi|^2 e^{-2L} \leq \sum_1^N \int_{C^N} \left|\frac{\partial \chi}{\partial \bar{z}_j}\right|^2 e^{-2L}$$

holds for all $\chi \in \Lambda_0^{0,0}(C^N)$ *if and only if* $c \leq \Sigma_1^N \lambda_j^-$.

Proof: First suppose $\lambda_j > 0$ for all j. Choose $\chi \in \Lambda_0^{0,0}(C^N)$ with $\chi(0) = 1$ and set $\chi^\epsilon(z) = \chi(\epsilon z)$. Then

$$\int |\chi^\epsilon|^2 e^{-2L} \to \int e^{-2L} < \infty \quad \text{as} \quad \epsilon \to 0 ,$$

but $\dfrac{\partial \chi^\epsilon}{\partial \bar{z}_j} = \epsilon \left(\dfrac{\partial \chi}{\partial \bar{z}_j}\right)^\epsilon$, so

$$\int \left|\frac{\partial \chi^\epsilon}{\partial \bar{z}_j}\right|^2 e^{-2L} = \mathcal{O}(\epsilon^2) \quad \text{as} \quad \epsilon \to 0 .$$

Thus we must have $c \leq 0 = \Sigma_1^N \lambda_j^-$. More generally, if the λ_j are all non-negative and precisely m of them are zero, then

$$\int |\chi^\epsilon|^2 e^{-2L} \sim \epsilon^{-m} \quad \text{and} \quad \int \left|\frac{\partial \chi^\epsilon}{\partial \bar{z}_j}\right|^2 e^{-2L} \sim \epsilon^{2-m} \quad \text{as} \quad \epsilon \to 0 ,$$

so again we must have $c \leq 0 = \Sigma_1^N \lambda_j^-$.

In general,

$$(3.2.15) \int \left|\frac{\partial \chi}{\partial \bar{z}_j}\right|^2 e^{-2L} = \int \frac{\partial \chi}{\partial \bar{z}_j} \frac{\partial \bar{\chi}}{\partial z_j} e^{-2L}$$

$$= -\int \chi \frac{\partial^2 \bar{\chi}}{\partial z_j \partial \bar{z}_j} e^{-2L} + 2\int \chi \frac{\partial \bar{\chi}}{\partial z_j} \lambda_j z_j e^{-2L}$$

$$= \int \frac{\partial \chi}{\partial z_j} \frac{\partial \bar{\chi}}{\partial \bar{z}_j} e^{-2L} - 2\int \chi \frac{\partial \bar{\chi}}{\partial \bar{z}_j} \lambda_j \bar{z}_j e^{-2L}$$

$$-2\int \frac{\partial \chi}{\partial z_j} \bar{\chi} \lambda_j z_j e^{-2L} - 2\int |\chi|^2 \lambda_j e^{-2L}$$

$$+4\int |\chi|^2 \lambda_j^2 |z_j|^2 e^{-2L}$$

$$= \int \left|\frac{\partial \chi}{\partial z_j} - 2\chi \lambda_j \bar{z}_j\right|^2 e^{-2L} - 2\lambda_j \int |\chi|^2 e^{-2L} .$$

Now set $L' = \Sigma_1^N |\lambda_j| \, |z_j|^2$, $L^- = \Sigma_1^N \lambda_j^- |z_j|^2$, and $\chi'(z_1, \ldots, z_n) = \chi(w_1, \ldots, w_n) e^{2L^-}$ where $w_j = z_j$ if $\lambda_j \geq 0$ and $w_j = \bar{z}_j$ if $\lambda_j < 0$. Then for $\lambda_j \geq 0$,

$$(3.2.16) \qquad \int \left|\frac{\partial \chi}{\partial \bar{z}_j}\right|^2 e^{-2L} = \int \left|\frac{\partial \chi'}{\partial \bar{z}_j}\right|^2 e^{-4L^-} e^{-2L} = \int \left|\frac{\partial \chi'}{\partial \bar{z}_j}\right|^2 e^{-2L'},$$

and for $\lambda_j < 0$, by (3.2.15),

$$(3.2.17) \qquad \int \left|\frac{\partial \chi}{\partial \bar{z}_j}\right|^2 e^{-2L} = \int \left|\frac{\partial \chi}{\partial z_j} - 2\chi\lambda_j \bar{z}_j\right|^2 e^{-2L} + 2\lambda_j^- \int |\chi|^2 e^{-2L}$$

$$= \int \left|\frac{\partial \chi'}{\partial \bar{z}_j} e^{-2L^-} + 2\chi'\lambda_j \bar{z}_j e^{-2L^-} - 2\chi'\lambda_j \bar{z}_j e^{-2L^-}\right|^2 e^{-2L}$$

$$+ 2\lambda_j^- \int |\chi'|^2 e^{-4L^-} e^{-2L}$$

$$= \int \left|\frac{\partial \chi'}{\partial \bar{z}_j}\right|^2 e^{-2L'} + 2\lambda_j^- \int |\chi'|^2 e^{-2L'}.$$

Combining (3.2.16) and (3.2.17), therefore, the inequality

$$2c \int |\chi|^2 e^{-2L} \leq \sum_1^N \left|\frac{\partial \chi}{\partial \bar{z}_j}\right|^2 e^{-2L} \quad \text{for all} \quad \chi \in \Lambda_0^{0,0}(C^N)$$

is equivalent to the inequality

$$(3.2.18) \qquad 2\left(c - \sum_1^N \lambda_j^-\right) \int |\chi'|^2 e^{-2L'} \leq \sum_1^N \left|\frac{\partial \chi'}{\partial \bar{z}_j}\right|^2 e^{-2L'}$$

for all $\chi' \in \Lambda_0^{0,0}(C^N)$. If $c \leq \Sigma_1^N \lambda_j^-$, (3.2.18) is clearly valid; conversely, by the result for nonnegative λ's applied to $\{|\lambda_j|\}_1^N$, (3.2.18) implies that $c - \Sigma_1^N \lambda_j^- \leq 0$. Q.E.D.

Back on our manifold M, given $p \in bM$ we choose a special boundary chart U around p so that the Levi form is diagonal at p with eigenvalues $\lambda_1, \ldots, \lambda_{n-1}$. Let $\{z_j = x_j + iy_j\}_1^n$ be complex coordinates on U

with origin at p (we henceforth write $p = 0$) such that $\frac{\partial}{\partial z_j}\big|_0 = -L_j\big|_0$ and $dy_n\big|_0 = \operatorname{Im} dz_n\big|_0 = -dr\big|_0$. Then in a sufficiently small neighborhood of 0, bM is described by the equation $y_n = \rho(x,y')$ where $y' = (y_1,\dots,y_{n-1})$; thus $r = (\rho - y_n)(1 + \mathcal{O}(|z|))$. ρ has a zero of order 2 at 0, so by Taylor's theorem,

(3.2.19) $$\rho = \sum_1^{n-1} \lambda_j |z_j|^2 + \operatorname{Re} A(z') + \mathcal{O}(|z'|\,|z_n| + |z_n|^2 + |z|^3)$$

where $z' = (z_1,\dots,z_{n-1})$ and $A(z')$ is a homogeneous holomorphic polynomial of degree 2.

(3.2.20) LEMMA. *If the basic estimate holds for all* $\phi \in \mathfrak{D}^{p,q} \cap \Lambda_0^{p,q}(U \cap \bar M)$ *for some neighborhood* U *of* 0, *then there exist constants* $c, c_1 > 0$ *such that for every* $\delta > 0$ *there is a neighborhood* $V \subset U$ *of* 0 *with*

$$c\int_{bM} |\phi|^2 \le (1+\delta)\sum_{kIJ} \left\|\frac{\partial \phi_{IJ}}{\partial \bar z_k}\right\|^2 + \sum_{IJ}\sum_{k\epsilon J} \lambda_k \int_{bM} |\phi_{IJ}|^2 + c_1\|\phi\|^2$$

uniformly for $\phi \in \mathfrak{D}^{p,q} \cap \Lambda_0^{p,q}(V \cap \bar M)$.

Proof: From Lemma (3.2.3) we see that the basic estimate implies this inequality with $\delta = 0$ and $\frac{\partial}{\partial \bar z_k}$ replaced by $\bar L_k$ for sufficiently small V. But since $\frac{\partial}{\partial \bar z_k}\big|_0 = -\bar L_k\big|_0$,

$$\left|\left\|\frac{\partial \phi_{IJ}}{\partial \bar z_k}\right\|^2 - \|\bar L_k \phi_{IJ}\|^2\right| \le \delta \sum_j \left\|\frac{\partial \phi_{IJ}}{\partial \bar z_j}\right\|^2$$

where δ can be made arbitrarily small by requiring supp ϕ to be small. Q.E.D.

(3.2.21) THEOREM. *If the basic estimate holds in* $\mathfrak{D}^{p,q}$, *then* M *satisfies condition* Z(q).

Proof: For each $p \in bM$, we shall show that the Levi form at p has $n-q$ positive eigenvalues or $q+1$ negative eigenvalues at p by working in a special boundary chart V near p on which the conclusion of Lemma (3.2.20) holds. First we prove the theorem for functions, i.e., $p = q = 0$, for which $\mathcal{D}^{0,0} = \Lambda^{0,0}(\overline{M})$. Fix $\psi \in \Lambda_0^{0,0}(V \cap \overline{M})$ and set $f^t(z) = \psi(tz)e^{it^2 z_n}$. Then

$$|f^t(z)|^2 = |\psi(tz)|^2 e^{-2t^2 y_n}, \tag{3.2.22}$$

$$\left|\frac{\partial f^t}{\partial \bar{z}_k}(z)\right|^2 = t^2 \left|\frac{\partial \psi}{\partial \bar{z}_k}(tz)\right|^2 e^{-2t^2 y_n}, \tag{3.2.23}$$

$$\int_{bM} |f^t(z)|^2 = \int_{bM} |\psi(tz)|^2 e^{-2t^2 y_n}(a(z)\,dx\,dy' + \ldots) \tag{3.2.24}$$

where $a(0) = 1$ and the dots denote components of the volume element on bM whose coefficients vanish at 0. (As before, $y = (y_1, \ldots, y_{n-1})$, and likewise for x' and z'.) We now make the change of variables $Y_n = t^2 y_n$, $Y' = ty'$, $X = tx$; hence also $Z' = tz'$. Then from (3.2.24),

$$\int_{bM} |f^t(z)|^2 = \int_{bM} \left|\psi\left(X, Y', \frac{Y_n}{t}\right)\right|^2 e^{-2Y_n} t^{1-2n}\left(a\left(\frac{X}{t}, \frac{Y'}{t}, \frac{Y_n}{t^2}\right) dX dY' + \mathcal{O}\left(\frac{1}{t}\right)\right),$$

so

$$t^{2n-1}\int_{bM} |f^t(z)|^2 = \int_{bM} \left|\psi\left(X, Y', \frac{Y_n}{t}\right)\right|^2 e^{-2Y_n} a\left(\frac{X}{t}, \frac{Y'}{t}, \frac{Y_n}{t^2}\right) dX\,dY' + \mathcal{O}\left(\frac{1}{t}\right).$$

Using (3.2.19) and setting $L(Z') = \Sigma_1^{n-1}\lambda_j|Z_j|^2$, we obtain

$$t^{2n-1}\int_{bM} |f^t(z)|^2 = \int_{bM} \left|\psi\left(X, Y', \frac{Y_n}{t}\right)\right|^2 a\left(\frac{X}{t}, \frac{Y'}{t}, \frac{Y_n}{t^2}\right) \exp\left[-2L(Z') - 2\mathrm{Re}\,A(Z') + \mathcal{O}\left(\frac{1}{t}\right)\right] dX dY' + \mathcal{O}\left(\frac{1}{t}\right)$$

which converges to

$$\int_{R^{2n-1}} |\psi(X, Y', 0)|^2 \exp(-2L(Z') - 2\mathrm{Re}\,A(Z'))\,dX dY'$$

as $t \to \infty$. Moreover, as $t \to \infty$, we see from (3.2.23) that

$$t^{2n-1}\left\|\frac{\partial f^t}{\partial \bar z_j}\right\|^2 = t^{2n+1}\int_M \left|\frac{\partial \psi}{\partial \bar z_j}(tz)\right|^2 e^{-2t^2 y_n}\,dx\,dy$$

$$= \int_{R^{2n-1}}\int_{\rho(X,Y')}^{\infty} \left|\frac{\partial \psi}{\partial \bar z_j}(X',Y',\frac{Y^n}{t})\right|^2 e^{-2Y_n} d\,Y_n\,dX\,dY'$$

$$\to \int_{R^{2n-1}}\left[\int_{\rho(X,Y')}^{\infty} e^{-2Y_n} d\,Y_n\right]\left|\frac{\partial \psi}{\partial \bar z_j}(X,Y',0)\right| dX dY'$$

$$= \tfrac{1}{2}\int_{R^{2n-1}} \left|\frac{\partial \psi}{\partial \bar z_j}(X,Y',0)\right| \exp(-2L(Z') - 2\mathrm{Re}\,A(Z'))dX\,dY',$$

and also, from (3.2.22),

$$t^{2n-1}\|f^t\|^2 = t^{2n-1}\int_M |\psi(tz)|^2 e^{-2t^2 y_n}\,dx\,dy$$

$$= \frac{1}{t^2}\int_{R^{2n-1}}\int_{\rho(X,Y')}^{\infty} \left|\psi(X,Y',\frac{Y_n}{t})\right|^2 e^{-2Y_n} dY_n\,dX\,dY'$$

$$\to 0\,.$$

Applying Lemma (3.2.20) for $p = q = 0$ to f^t and letting $t \to \infty$, therefore, we obtain

(3.2.25) $$c\int_{R^{2n-1}} |\psi(X,Y',0)|^2 \exp(-2L(Z') - 2\mathrm{Re}\,A(Z'))\,dX\,dY'$$

$$\le \frac{(1+\delta)}{2}\sum_1^n \int_{R^{2n-1}} \left|\frac{\partial \psi}{\partial \bar z_k}(X,Y',0)\right|^2 \exp(-2L(Z') - 2\mathrm{Re}\,A(Z'))\,dX\,dY'.$$

Let us now take ψ to be of the form $\psi(z) = \psi_1(z')e^{A(z')}\psi_2(z_n)$ where $\frac{\partial \psi_2}{\partial \bar z_n}\Big|_{y_n=0} = 0$. This condition does not restrict the values of ψ_2 on the hyperplane $y_n = 0$, so we may also arrange that $\int_R |\psi_2(x_n,0)|^2 dx_n = 1$. Integrating (3.2.25) with respect to X_n, we find that

$$c\int_{R^{2n-2}} |\psi_1(Z')|^2 e^{-2L(Z')} dX'dY' \le \frac{(1+\delta)}{2}\sum_1^{n-1}\int_{R^{2n-2}} \left|\frac{\partial\psi_1}{\partial\bar{z}_k}(Z')\right|^2 e^{-2L(Z')} dX'dY'$$

In other words,

$$\frac{2c}{1+\delta}\int_{C^{n-1}} |\psi_1|^2 e^{-2L} \le \sum_1^{n-1}\int_{C^{n-1}} \left|\frac{\partial\psi_1}{\partial\bar{z}_k}\right|^2 e^{-2L}$$

for all $\psi_1 \epsilon \Lambda_0^{0,0}(C^{n-1})$. Lemma (3.2.14) therefore implies that $0 < \frac{c}{1+\delta} \le \Sigma_1^{n-1}\lambda_j^-$, so at least one λ_j must be negative. But this is condition $Z(0)$, so the theorem is proved for $p = q = 0$.

The proof for general p and q now follows easily. Let I be an arbitrary multi-index of length p, and J a multi-index of length q with $n \notin J$. Define $\phi^t \epsilon \mathcal{D}^{p,q} \cap \Lambda_0^{p,q}(V \cap \bar{M})$ by $\phi^t_{IJ} = f^t$, $\phi^t_{I'J'} = 0$ if $I'J' \ne IJ$, where f^t is as above. Then the preceding argument goes through without change, except that when we apply Lemma (3.2.20) we obtain the constant $c - \Sigma_{k\epsilon J}\lambda_k$ instead of c on the left. The inequality implied by Lemma (3.2.14) is then $\frac{1}{1+\delta}(c - \Sigma_{k\epsilon J}\lambda_k) \le \Sigma_1^{n-1}\lambda_j^-$. Since δ is arbitrarily small, we have $0 < c \le \Sigma_{k\epsilon J}\lambda_k + \Sigma_1^{n-1}\lambda_j^-$. But if at most $n-q-1$ of the λ_j's were positive and at most q were negative, we could choose J so that $k \epsilon J$ if $\lambda_k < 0$ and $k \notin J$ if $\lambda_k > 0$, and this inequality would be false. Q.E.D.

(3.2.26) COROLLARY. *The basic estimate holds in* $\mathcal{D}^{p,0}$ *if and only if the Levi form has at least one negative eigenvalue at each point. In this case, if* M *is connected the holomorphic functions on* M *are all constant.*

Proof: The holomorphic functions on M which are smooth up to the boundary constitute the space $\mathcal{H}^{0,0}$, which is finite-dimensional by Proposition (3.1.11). If $f \epsilon \mathcal{H}^{0,0}$ were non-constant, by connectedness we would have $df \ne 0$ and all the powers f^j of f would be linearly independent elements of $\mathcal{H}^{0,0}$, which is impossible. To obtain the result for arbitrary holomorphic f, we shrink M a little bit to produce a domain M_0 which still satisfies condition $Z(0)$ and on which f is smooth up to the boundary. Then f is constant on M_0, hence on M. Q.E.D.

CHAPTER IV

APPLICATIONS

In this chapter we shall present a few salient applications of the $\bar{\partial}$-Neumann problem. This list, however, is not complete, and we shall see other applications in later chapters.

1. *The Newlander-Nirenberg theorem*

We shall prove this theorem by deforming the given almost-complex structure into a "flat" one. Let $M \subset M'$ be a smooth, compact, 2n-dimensional real manifold with boundary. Suppose that for each t, $0 \leq t \leq 1$, we are given an integrable almost-complex structure on M' with Hermitian metric, varying smoothly in t, which we shall denote by M'_t. (In general, we shall identify all the analytical objects belonging to M'_t by affixing a subscript t.) Note that if M_0 satisfies condition $Z(q)$ then so does M_t for small t, since the eigenvalues of the Levi form are continuous in t. Since the Sobolev norms are independent of t up to equivalence, we shall always use the norms given by the metric at $t=0$.

(4.1.1) LEMMA. *Suppose* M_0 *is strongly pseudoconvex. Then for sufficiently small* t, $H_0\colon \mathcal{H}_t^{0,1} \to \mathcal{H}_0^{0,1}$ *is injective; in particular,* $\dim \mathcal{H}_t^{0,1} \leq \dim \mathcal{H}_0^{0,1}$

Proof: For sufficiently small t, the smallest eigenvalue of the Levi form is bounded away from zero, so the proofs of Theorems (2.4.4) and (3.2.10) show that the estimates $|||D\phi|||_{-1/2} \lesssim E_t(\phi)$ and $E_t(\phi)^2 \lesssim Q_t(\phi,\phi)$ hold uniformly for small t. Suppose that there were a sequence $t_j \to 0$ and $\phi_j \in \mathcal{H}_{t_j}^{0,1}$ with $\|\phi_j\| = 1$ and $H_0\phi_j = 0$. Then $Q_{t_j}(\phi_j,\phi_j) = (\phi_j,\phi_j)_{t_j} \leq$ const., so by compactness of the norm $|||D(\cdot)|||_{-1/2}$ (cf. Appendix, §3) (and passage to a subsequence if necessary) we see that ϕ_j converges to some $\phi_0 \in H_0^{p,q}$ satisfying $\|\phi_0\| = 1$ and $H_0\phi_0 = \lim H_0\phi_j = 0$. However, for all $u \in \mathcal{D}_0^{0,0}$, we have

$$(\phi_0,\bar{\partial}_0 u) = \lim(\phi_j,\bar{\partial}_0 u) = \lim[(\phi_j,(\bar{\partial}_0 - \bar{\partial}_{t_j})u) + (\phi_j,\bar{\partial}_{t_j}u)]$$
$$= \lim(\phi_j,(\bar{\partial}_0 - \bar{\partial}_{t_j})u) = 0$$

since $\phi_j \in \mathcal{H}_{t_j}^{0,1}$, and for all $\psi \in \mathcal{D}_0^{0,2}$,

$$(\phi_0,\vartheta_0\psi) = \lim(\phi_j,\vartheta_0\psi) = \lim[(\phi_j,(\vartheta_0 - \vartheta_{t_j})\psi) + (\phi_j,\vartheta_{t_j}\psi)]$$
$$= \lim(\phi_j,(\vartheta_0 - \vartheta_{t_j})\psi) + \lim\int_{bM} \langle\phi_j,\sigma(\vartheta_{t_j},dr)\psi\rangle = 0\,.$$

Thus $\phi_0 \perp \text{Range}(\bar{\partial}_0) \oplus \text{Range}(\bar{\partial}_0^*)$, so $\phi_0 \in \mathcal{H}_0^{0,1}$. Contradiction. Q.E.D.

(4.1.2) LEMMA. *Suppose* M_0 *is strongly pseudoconvex and* $\mathcal{H}_0^{0,1} = 0$. *Then the estimate* $\|N_t\alpha\|_{s+1} \lesssim \|\alpha\|_s$ *for* $\alpha \in \Lambda^{0,1}(\bar{M})$ *holds uniformly for small* t.

Proof: As in the proof of Lemma (4.1.1), the estimate $E_t(\phi)^2 \lesssim Q_t(\phi,\phi)$ holds uniformly for small t. Hence the Main Theorem yields the estimate $\|\phi\|_{s+1} \lesssim \|F_t\phi\|_s$ uniformly in t. By Lemma (4.1.1), $\mathcal{H}_t^{0,1} = 0$ for small t, so $F_tN_t\alpha = \alpha + N_t\alpha$ for all $\alpha \in \Lambda^{0,1}(\bar{M})$. Hence $\|N_t\alpha\|_{s+1} \lesssim \|\alpha\|_s + \|N_t\alpha\|_s$ uniformly in t, so by induction on s it suffices to show that $\|N_t\alpha\| \lesssim \|\alpha\|$ uniformly in t. Suppose to the contrary that there were sequences $t_j \to 0$, $\alpha_j \in \Lambda^{0,1}(\bar{M})$ with $\|N_{t_j}\alpha_j\| > j$ and $\|\alpha_j\| = 1$. Set $\phi_j = N_{t_j}\alpha_j$ and $\theta_j = \phi_j/\|\phi_j\|$. Then $Q_{t_j}(\theta_j,\theta_j) = (\frac{\alpha_j}{\|\phi_j\|} + \theta_j,\theta_j) \le 2$, so as in Lemma (4.1.1), passing to a subsequence we have $\theta_j \to \theta_0$ in $H_0^{0,1}$ and $\|\theta_0\| = 1$. But since $\square_{t_j}\theta_j = \frac{\alpha_j}{\|\phi_j\|} \to 0$, we see by the same argument as in Lemma (4.1.1) that $\theta_0 \perp \text{Range}(\square_{F_0})$. Hence $\theta_0 \in \mathcal{H}_0^{0,1} = 0$, which is a contradiction. Q.E.D.

(4.1.3) THEOREM (Newlander-Nirenberg). *An integrable almost-complex manifold is complex.*

Proof: Let X be an integrable almost-complex manifold of dimension $2n$. It suffices to show that for every $p \in X$ there is a neighborhood V of p and functions $z_1, \ldots, z_n$ on V with $\bar\partial z_j = 0$ for each j and $\{dz_j\}_1^n$ linearly independent. The z_j's then form a complex coordinate system on V, and if $w_1, \ldots, w_n$ are another such coordinate system on V', on $V \cap V'$ we have $\frac{\partial z_j}{\partial \bar w_k} = 0$ since $\bar\partial z_j = 0$. Thus the transition functions are holomorphic, so we have defined a complex structure on X.

Therefore let U be a real coordinate patch with origin at p and coordinates $x_1, \ldots, x_{2n}$, and let the almost-complex structure be given on U by $\Pi_{1,0} dx_k = \Sigma_1^{2n} a_k^j(x)\, dx_j$. By renumbering the coordinates, we may assume that $\{\Sigma_1^{2n} a_k^j(0)\, dx_j\}_{k=1}^n$ are linearly independent. Then $z_{k,0} = \Sigma_1^{2n} a_k^j(0)\, x_j$ form a coordinate system defining a complex structure on U. We then define a family U_t of integrable almost-complex structures on U by $(\Pi_{0,1})_t\, dx_k = \Sigma_1^{2n} a_k^j(tx)\, dx_j$ for $0 \le t \le 1$, and we choose Hermitian metrics on U_t varying smoothly in t, the metric on U_0 being the flat metric $\langle dz_{i,0}, dz_{j,0}\rangle = 2\delta_{ij}$. Then U_1 is the original structure and U_0 is complex. Let $M = \{x \in U : \Sigma_1^n |z_{k,0}(x)|^2 < 1\}$; then bM is defined by $r = \Sigma_1^n |z_{k,0}(x)|^2 - 1$, so $\langle \partial_0 \bar\partial_0 r, dz_{k,0} \wedge d\bar z_{k,0}\rangle = 1$ for all k and hence M_0 is strongly pseudoconvex. Moreover, $\mathcal{H}_0^{0,1} = 0$ by Corollary (2.1.5). Set $z_{k,t} = H_t z_{k,0}$. For sufficiently small t, M_t is strongly pseudoconvex, and hence by Proposition (3.1.17), Corollary (3.2.12), and the fact that $\mathfrak{D}_0^{0,0} = \Lambda^{0,0}(\bar M)$, $z_{k,t} = z_{k,0} - \vartheta_t N_t \bar\partial_t z_{k,0}$. Then by Lemma (4.1.2), $\|z_{k,t} - z_{k,0}\|_s = \|\vartheta_t N_t \bar\partial_t z_{k,0}\|_s \lesssim \|N_t \bar\partial_t z_{k,0}\|_{s+1} \lesssim \|\bar\partial_t z_{k,0}\|_s$ uniformly in t for small t. But by our construction, $\bar\partial_t z_{k,0} \to \bar\partial_0 z_{k,0} = 0$ smoothly as $t \to 0$, which implies that $\{dz_{k,t}\}_1^n$ are linearly independent for small t.

In short, for some sufficiently small $t_0 > 0$, $\{z_{k,t_0}\}_1^n$ is a holomorphic coordinate system on M_{t_0}. But now we are done: set $V = t_0 M =$

$\{x \in U: \Sigma_1^n |z_{k,0}(x)|^2 < t_0\}$, and define $z_k(x) = z_{k,t_0}(\frac{x}{t_0})$. Then $\{z_k\}_1^n$ is a coordinate system in V which is holomorphic with respect to the original almost-complex structure. Q.E.D.

Remark. The original proof of Newlander and Nirenberg [35] was based on reducing the given problem to a non-linear problem. The present proof deals directly with the linear problem. Another proof, which applies to general elliptic transitive pseudogroup structures and which uses less heavy machinery than our Main Theorem, has been discovered by Malgrange [32]. A generalization in a different direction, extending the classical Frobenius theorem to complex vector fields, is known as the Frobenius-Nirenberg theorem; cf. Nirenberg [36a], Hörmander [17].

2. *The Levi problem*

In 1958 H. Grauert [13] solved a long-standing problem in complex analysis by showing, using sheaf theory, that every strongly pseudoconvex manifold is holomorphically convex. In this section we shall prove an equivalent theorem, which we already discussed briefly in §1.1, by means of the $\bar{\partial}$-Neumann problem. More precisely, we say that the compact complex manifold with boundary M is a *domain of holomorphy* if for every $p \in bM$ there is a holomorphic function on M with a singularity at p. (For regions in $\mathbf{C}^n$ there are several equivalent definitions of a domain of holomorphy, and this is one of them; see, e.g., Bers [6].) We shall prove:

(4.2.1) THEOREM. *If* M *is strongly pseudoconvex, then* M *is a domain of holomorphy.*

As we indicated in §1.1, the proof depends on the construction for each $p \in bM$ of a holomorphic function in a neighborhood of p which blows up at p. To accomplish this, let $z_1, \ldots, z_n$ be a coordinate system in a neighborhood U of p with origin at p such that $dz_1|_0, \ldots, dz_{n-1}|_0$

are tangent to bM and $dr|_0 = \operatorname{Re} dz_n|_0$. The Levi form is then given by the matrix $2\left(\frac{\partial^2 r}{\partial z_i \partial \bar z_j}\right)_{i,j=1}^{n-1}$. The *Levi polynomial* on U is the second-degree holomorphic polynomial

$$\Lambda(z) = z_n + 2\sum_{i,j=1}^{n-1} \frac{\partial^2 r}{\partial z_i \partial z_j}(0) z_i z_j \,.$$

(4.2.2) LEMMA. *If the Levi form at* p *is positive definite, then there exists a neighborhood* $V \subset U$ *of* p *such that for all sufficiently small* $a > 0$, $r > \frac{1}{2}a$ *on the intersection of* V *with the hypersurface* $\Lambda(z) = a$ *and* $r > 0$ *on the intersection of* $V - \{p\}$ *with the hypersurface* $\Lambda(z) = 0$.

Proof: By Taylor's theorem, since r is real we have in our special coordinate system $\{z_i\}$,

$$r(z) = \operatorname{Re} z_n + 2\operatorname{Re}\sum_{i,j=1}^{n-1} \frac{\partial^2 r}{\partial z_i \partial z_j}(0) z_i z_j + 2\sum_{i,j=1}^{n-1} \frac{\partial^2 r}{\partial z_i \partial \bar z_j}(0) z_i \bar z_j + \mathcal{O}(|z|^3)\,.$$

On the hypersurface $\Lambda(z) = a$, $z_n = a - 2\sum_{i,j=1}^{n-1} \frac{\partial^2 r}{\partial z_i \partial z_j}(0) z_i z_j$; substituting this value for z_n we obtain

$$r(z) = a + 2\sum_{i,j=1}^{n-1} \frac{\partial^2 r}{\partial z_i \partial \bar z_j}(0) z_i \bar z_j + \mathcal{O}(a^2) + \mathcal{O}(a|z|) + \mathcal{O}(|z|^3)\,.$$

By hypothesis, $2\sum_{i,j=1}^{n-1} \frac{\partial^2 r}{\partial z_i \partial \bar z_j}(0) z_i \bar z_j \geq c|z|^2$ for some $c > 0$. We can thus choose a neighborhood $V \subset U$ so small that for $z \in V$ the $\mathcal{O}(|z^3|)$ term is $\leq \frac{1}{2}c|z|^2$ in absolute value and the $\mathcal{O}(a|z|)$ term is $\leq \frac{1}{4}a$ in absolute value. We then take a to be small enough that the $\mathcal{O}(a^2)$ term is $\leq \frac{1}{4}a$ in absolute value. Under these circumstances, $r(z) \geq \frac{1}{2}a + \frac{1}{2}c|z|^2$. Since $\Lambda(0) = 0$, we have $r(z) > \frac{1}{2}a$ if $a > 0$ and $r(z) > 0$ if $a = 0$ and $z \neq 0$. Q.E.D.

Proof of Theorem (4.2.1): With V and Λ as in Lemma (4.2.2), choose $\zeta \in \Lambda_0^{0,0}(V)$ with $\zeta = 1$ on a neighborhood of $p = 0$. Set $f(z) = \zeta(z)/\Lambda(z)$ and extend f to be zero outside V; by Lemma (4.2.2), the set where $\Lambda(z)$ vanishes intersects $\overline{M} \cap V$ only at p and hence $f \in \Lambda^{0,0}(\overline{M} - \{p\})$. Next, set $\alpha = \bar{\partial} f$. Then $\alpha = 0$ except where $d\zeta \neq 0$ since Λ is holomorphic, and therefore $\alpha \in \Lambda^{0,1}(\overline{M})$. If $n \geq 3$ then $f \in H_0^{0,0}$, so α is in the range of the Hilbert space operator $\bar{\partial}$. For $n = 2$, set $f_a(z) = \zeta(z)/(\Lambda(z) - a)$ for small $a > 0$; then by Lemma (4.2.2) $f_a \in \Lambda^{p,q}(\overline{M})$, and $\bar{\partial} f_a$ converges to α pointwise as $a \to 0$, hence in $H_0^{0,1}$ since $\{\bar{\partial} f_a\}$ are uniformly bounded. Now the basic estimate holds in $\mathfrak{D}^{0,1}$; therefore $\alpha \in$ Range $(\bar{\partial})$ by Corollary (3.1.13).

By Proposition (3.1.15), $\alpha = \bar{\partial} g$ where $g \in \Lambda^{0,0}(\overline{M})$. Set $h = f - g$. Then $\bar{\partial} h = \alpha - \alpha = 0$, and h has the same singularity as $1/\Lambda$ at p. Q.E.D.

Remarks:

(1) We can construct a function with a singularity of arbitrarily high order at p by using large powers of Λ instead of Λ.

(2) There is a notion of pseudoconvexity for domains in C^n which does not depend on having a smooth boundary, and it is one of the oldest results in several complex variables that every domain of holomorphy in C^n is pseudoconvex. On the other hand, it is not hard to show that every pseudoconvex domain is the increasing limit of strongly pseudoconvex domains, and by a theorem of Behnke and Stein, the increasing limit of domains of holomorphy is a domain of holomorphy. Therefore: a region in C^n is a domain of holomorphy if and only if it is pseudoconvex. For these matters see, e.g., Bers [6].

(3) This theorem illustrates the ways in which the Main Theorem and its corollaries may break down if the basic estimate fails to hold. In the first place, if M is strongly pseudoconvex then the basic estimate does not hold in $\mathfrak{D}^{0,0}$. Theorem (4.2.1) yields a non-constant holomorphic function on M, and hence by the argument of Corollary (3.2.26), $\dim \mathcal{H}^{0,0} = \infty$. On the other hand, let M be a deformed region between two concentric spheres in C^n, thus:

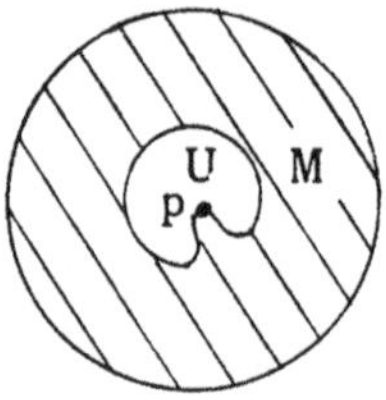

where the Levi form at p is positive definite. Here the basic estimate always fails (except for $q = n$) since the Levi form is negative definite on most of the inner boundary. We can construct a Levi polynomial Λ at p, but Theorem (4.2.1) breaks down since any holomorphic function on M can be extended holomorphically to the inner region U by Hartogs' theorem [18]. On the one hand, setting $f(z) = \zeta(z)/(\Lambda(z))^N$ and $f_a(z) = \zeta(z)/(\Lambda(z)-a)^N$ for N large enough so that $f(z) \notin H_0^{0,0}$, we have $\alpha = \bar{\partial} f = \lim_{a \to 0} \bar{\partial} f_a$, hence α is in the closure of Range $(\bar{\partial})$. But if $\alpha = \bar{\partial} g$ where $g \in H_0^{0,0}$, then $f - g$ would be holomorphic in M, hence extendable to U, hence square-integrable. Therefore Range $(\bar{\partial})$ is not closed. On the other hand, for $n \geq 3$, $f(z) = \zeta(z)/\Lambda(z)$ *is* square-integrable, so we conclude from the failure of Theorem (4.2.1) that the equation $\bar{\partial} g = \alpha$, $\alpha \in$ Range $(\bar{\partial}) \cap \Lambda^{0,1}(\bar{M})$, may not possess a smooth solution $g \in \Lambda^{0,0}(\bar{M})$.

(4) In case dim M = 2, Kohn [25] has recently shown that, using a more complicated polynomial, the proof of Theorem (4.2.1) remains valid under weaker pseudoconvexity hypotheses.

3. *Remarks on $\bar{\partial}$ cohomology*

There are several natural cohomology groups associated to the $\bar{\partial}$ complex on the Hermitian manifold with boundary M:

$$H^{(p,q)}(M) = \frac{\{\phi \in \Lambda^{p,q}(M) : \bar{\partial}\phi = 0\}}{\bar{\partial}\Lambda^{p,q-1}(M)},$$

$$H^{(p,q)}(\bar{M}) = \frac{\{\phi \in \Lambda^{p,q}(\bar{M}) : \bar{\partial}\phi = 0\}}{\bar{\partial}\Lambda^{p,q-1}(\bar{M})},$$

$$\tilde{H}^{(p,q)}(M) = \frac{\{\phi \in H_0^{p,q} \cap \mathrm{Dom}\,(\bar{\partial}) : \bar{\partial}\phi = 0\}}{\bar{\partial}(H_0^{p,q-1} \cap \mathrm{Dom}\,(\bar{\partial}))}.$$

From Theorem (3.1.14) we know that $H^{(p,q)}(\overline{M}) \cong \tilde{H}^{(p,q)}(M) \cong \mathcal{H}^{p,q}$ provided M satisfies condition Z(q). On the other hand, the Dolbeault theorem (cf. [40]) asserts that $H^{(p,q)}(M) \cong H^q(M;\Omega^p)$ where Ω^p is the sheaf of germs of (p,0)-forms. The problem of relating these important groups to the preceding ones is non-trivial; here we simply state the result:

(4.3.1) THEOREM. *If* M *satisfies conditions* Z(q) *and* Z(q+1) *then* $H^q(M;\Omega^p) \cong \mathcal{H}^{p,q}$.

For a proof the reader is referred to Hörmander [16], who uses weight functions to temper the behavior of forms at the boundary. He shows that if $a < 0$ is small enough so that $M_c = \{x \in M : r(x) < c\}$ also satisfies conditions Z(q) and Z(q+1) for $a \leq c \leq 0$, the restriction homomorphisms $H^{(p,q)}(M) \to H^{(p,q)}(\overline{M}_a)$ and $H^{(p,q)}(\overline{M}) \to H^{(p,q)}(\overline{M}_a)$ are isomorphisms, and the theorem follows immediately.

By now it is clear that the holomorphic degree p plays no role in the behavior of the $\bar{\partial}$ complex and that we could obtain the same results by replacing (p,q)-forms with V-valued (0,q)-forms where V is *any* holomorphic vector bundle. More specifically, let $\mathcal{A}^q$ denote the sheaf of germs of (0,q)-forms and $\mathcal{O}$ the sheaf of germs of sections of the holomorphic vector bundle V. We can then form a $\bar{\partial}$ complex out of the sections of the sheaves $\mathcal{O} \otimes \mathcal{A}^q$, $0 \leq q \leq n$, and take the cohomology groups

$$H^q_V(M) = \frac{\{\phi \in H^0(M;\mathcal{O}\otimes\mathcal{A}^q) : \bar{\partial}\phi = 0\}}{\bar{\partial}H^0(M;\mathcal{O}\otimes\mathcal{A}^{q-1})} .$$

(In particular, taking $\mathcal{O} = \Omega^p$, we recover the (p,q)-forms.) By the Dolbeault isomorphism, $H^q(M;\mathcal{O}) \cong H^q_V(M)$. Choosing a Hermitian metric on V, we can define the adjoint of $\bar{\partial}$ and set up the $\bar{\partial}$-Neumann problem. (For more details on this construction, see Kohn and Rossi [28].) The results of Chapters 2 and 3 then go through without any change. In particular, in view of the fact that locally free analytic sheaves are precisely those which arise as sections of holomorphic vector bundles, we have the following version of Cartan's Theorem B for strongly pseudoconvex manifolds:

(4.3.2) THEOREM. *If* M *is strongly pseudoconvex and* $\mathcal{O}$ *is any locally free analytic sheaf on* M, *then* $H^q(M;\mathcal{O})$ *is finite-dimensional for* $q > 0$.

As an application of sheaf cohomology, we now solve the Cousin problems, which generalize the theorems of Mittag-Leffler and Weierstrass for one complex variable.

Cousin Problem I. Let $U = \{U_i\}$ be an open covering of M, and for each i let F_i be a meromorphic function on U_i such that $F_i - F_j$ is holomorphic on $U_i \cap U_j$. Find a meromorphic function F on M such that $F - F_i$ is holomorphic on U_i.

Let $f_{ij} = F_i - F_j$ on $U_i \cap U_j$. If we can find holomorphic functions f_i on U_i such that $f_i - f_j = f_{ij}$ on $U_i - U_j$, we will be done, for the function F given by $F = F_i - f_i$ on U_i is globally well-defined and solves the problem. But the collection $\{f_{ij}\}$ defines a one-cocycle of the sheaf of germs of holomorphic functions $(= \Omega^0)$ relative to the covering $\mathfrak{U}$, and the existence of $\{f_i\}$ means that $\{f_{ij}\}$ is a coboundary. Therefore, using the Čech construction of sheaf cohomology, we see that Cousin I is always solvable on M if and only if $H^1(M;\Omega^0) = 0$. In particular, by virtue of Theorem (4.3.1) and Corollary (2.1.5), Cousin I is always solvable if M is a strongly pseudoconvex domain in $\mathbf{C}^n$.

Cousin Problem II. Let $\mathfrak{U} = \{U_i\}$ be an open covering of M, and for each i let F_i be a holomorphic function on U_i such that F_i/F_j is holomorphic on $U_i \cap U_j$. Find a holomorphic function F on M such that F/F_i is holomorphic on U_i.

As above, we set $f_{ij} = F_i/F_j$, which is a nonvanishing holomorphic function on $U_i \cap U_j$ (since F_j/F_i is also holomorphic). Then $\{f_{ij}\}$ is a one-cocycle of the sheaf $\mathcal{O}^*$ of nonvanishing holomorphic functions relative to the covering $\mathfrak{U}$, and it suffices to show that $\{f_{ij}\}$ is a coboundary. In other words, Cousin II is always solvable if and only if $H^1(M;\mathcal{O}^*) = 0$. We may formulate a topological obstruction to the solution by considering the exact sheaf sequence

$$0 \longrightarrow Z \longrightarrow \Omega^0 \xrightarrow{e^{2\pi i}} \mathcal{O}^* \longrightarrow 0 ,$$

which gives rise to the exact cohomology sequence

$$\ldots \to H^1(M;\Omega^0) \to H^1(M;\mathcal{O}^*) \to H^2(M;Z) \to H^2(M;\Omega^0) \to \ldots .$$

From this we see that Cousin II is always solvable on M if Cousin I is always solvable and $H^2(M;Z) = 0$. Moreover, if $H^2(M;\Omega^0) = 0$ (which is the case, for example, in Stein manifolds), the condition $H^2(M;Z) = 0$ is necessary, for then the map $H^1(M;\mathcal{O}^*) \to H^2(M;Z)$ is surjective.

Further information may be found in [6] and [18].

4. *Multiplier operators on holomorphic functions*

Let D be the unit disc in C^1, $L_2^N(D)$ the space of C^N-valued square-integrable functions on D, $H_2^N(D)$ the (closed) subspace of square-integrable holomorphic functions, and $H: L_2^N(D) \to H_2^N(D)$ the orthogonal projection. Let s be a smooth $(N \times N)$ matrix-valued function on D which is continuous on $\bar{D}$, and let $S_0 : L_2^N(D) \to L_2^N(D)$ be the operation of multiplication by s and S_1 its restriction to $H_2^N(D)$. It is a well-known result of Gohberg and Krein [12a] that if $\det s \neq 0$ on $T^1 = bD$, then the operator $S = HS_1 : H_2^N(D) \to H_2^N(D)$ is Fredholm, and its index is the degree of the map $\dfrac{\det s}{|\det s|} : T^1 \to T^1$. In this section we prove a theorem of U. Venugopalkrishna [45] which generalizes this result to strongly pseudoconvex manifolds.

If M is a strongly pseudoconvex manifold, we define $L_2^N(M)$, $H_2^N(M)$, H, S_0, S_1, and S as above, and we denote by $\Lambda_2^N(M)$ the space of square-integrable C^N-valued (0,1)-forms on M. (Thus $L_2^1(M) = H_0^{0,0}$, $H_2^1(M) = \mathcal{H}^{0,0}$, $\Lambda_2^1(M) = H_0^{0,1}$, and H is the harmonic projection.) We remark that the theory of the $\bar{\partial}$-Neumann problem goes through without change for vector-valued functions and forms, according to the remarks in the preceding section.

(4.4.1) THEOREM. *If* $\det s \neq 0$ *on* bM, *then* S *is Fredholm.*

The proof will be achieved by a series of lemmas, in which we do not assume $\det s \neq 0$.

(4.4.2) LEMMA. $\vartheta N : \Lambda_2^N(M) \to L_2^N(M)$ *is compact.*

Proof: If $\phi \in \Lambda_2^N(M)$, by Theorem (3.1.14) we have $\phi = \bar{\partial}\vartheta N\phi + \vartheta\bar{\partial}N\phi + H\phi$, and

$$\|\vartheta N\phi\|^2 = (\bar{\partial}\vartheta N\phi, N\phi) = (\phi, N\phi) - (\vartheta\bar{\partial}N\phi, N\phi) \tag{4.4.3}$$
$$= (\phi, N\phi) - (\bar{\partial}N\phi, \bar{\partial}N\phi) \leq (\phi, N\phi)$$

since $(H\phi, N\phi) = 0$. This shows that ϑN is bounded. Moreover, if $\{\phi_i\}$ is a sequence in $\Lambda_2^N(M)$ with $\|\phi_i\| \leq 1$, by passing to a subsequence we may assume that $\{N\phi_i\}$ is Cauchy since N is compact by Theorem (3.1.14). But then, from (4.4.3),

$$\|\vartheta N(\phi_i - \phi_j)\|^2 \leq (\phi_i - \phi_j, N(\phi_i - \phi_j)) \leq 2\|N(\phi_i - \phi_j)\|,$$

so $\{\vartheta N\phi_i\}$ is also Cauchy. Thus ϑN is compact. Q.E.D.

(4.4.4) LEMMA. $(I - H)S_1 : H_2^N(M) \to L_2^N(M)$ *is compact.*

Proof: If $f \in H_2^N(M)$ then $S_1 f \in \mathrm{Dom}\,(\bar{\partial})$ and $\bar{\partial}S_1 f = (\bar{\partial}s)f$. Thus $\bar{\partial}S_1$ is a bounded operator on $H_2^N(M)$; moreover, by Proposition (3.1.17), $(I - H)S_1 f = (\vartheta N)(\bar{\partial}S_1)f$. Since ϑN is compact by Lemma (4.4.2), we are done. Q.E.D.

(4.4.5) LEMMA. *If* $s = 0$ *on* bM *then* S *is compact.*

Proof: First suppose $K = \mathrm{supp}\ s$ is a compact subset of M. Let $f \in H_2^N(M)$. Since f is analytic, the derivatives of f on K are uniformly bounded by $\|f\|$ (to see this, apply the Cauchy integral formula; cf. [18].) By the Arzela-Ascoli theorem, then, $\{sf : f \in H_2^N(M), \|f\| \leq 1\}$ is relatively compact in the uniform topology, hence in $L_2^N(M)$, and the lemma follows.

In general, choose a sequence of multipliers $\{s_k\}$ with compact support in M such that $s_k \to s$ uniformly in $\bar{M}$. Then $\|s_k f - sf\|^2 \leq (\sup_M |s_k - s|^2)$ $\|f\|^2$, so $S_k \to S$ in the norm topology. Since each S_k is compact, so is S. Q.E.D.

Proof of Theorem (4.4.1). If $\det s \neq 0$ on bM, we can choose a multiplier t such that $st = ts = 1$ on bM. Let $r = st$, and denote the operators given by r and t by the corresponding capital letters as before. As operators on $L_2^N(M)$,

$$I - HS_0 HT_0 = I - HS_0 T_0 + HS_0(I-H)T_0 \,.$$

Restricting to $H_2^N(M)$,

$$I - HS_1 HT_1 = H(I-R_1) + HS_0(I-H)T_1 \,,$$

or

$$ST = I - H(I-R_1) - HS_0(I-H)T_1 \,.$$

The second term on the right is compact by Lemma (4.4.5) since $r = 1$ on bM, and the third term is compact by Lemma (4.4.4). Likewise $TS =$ $I +$ compact. Thus S is invertible modulo compact operators, i.e., S is Fredholm. Q.E.D.

Venugopalkrishna [45] goes on to prove the invariance of the index $\operatorname{ind} S = \dim \mathfrak{N}(S) - \operatorname{codim} \operatorname{Range}(S)$ under homotopies of s and that $\operatorname{ind}(ST) = \operatorname{ind} S + \operatorname{ind} T$, $\operatorname{ind}(S^*) = -\operatorname{ind} S$. He also computes ind S in the following cases:

(1) If $N = 1$ and bM is simply connected, then $\operatorname{ind} S = 0$.

(2) If M is the unit ball in $\mathbf{C}^n$, then $s : bM = S^{2n-1} \to GL(N, \mathbf{C})$ determines an element of the homotopy group $\pi_{2n-1}(GL(N,\mathbf{C}))$. By the Bott periodicity theorem, $\pi_{2n-1}(GL(N,\mathbf{C})) \cong \mathbf{Z}$ for $N \geq n$. For $M =$ unit ball and $N \geq n$, then, ind S is the integer given by the Bott isomorphism.

CHAPTER V

THE BOUNDARY COMPLEX

In this chapter we shall study the behavior of the boundary values of forms associated to the $\bar{\partial}$-Neumann problem. First, however, we shall make some remarks about some duality relations pertaining to the $\bar{\partial}$-Neumann problem. For the purposes of this chapter we shall make a slight change of notation and denote $\Lambda^{p,q}(\overline{M})$ by $\mathcal{A}^{p,q}$.

1. *Duality theorems*

Let $\mathcal{C}^{p,q} = \{\phi \in \mathcal{A}^{p,q} : \bar{\partial} r \wedge \phi = 0 \text{ on } bM\}$. Since $\sigma(\bar{\partial}, dr) = \bar{\partial} r \wedge (\cdot)$, we may also write

$$\mathcal{C}^{p,q} = \{\phi \in \mathcal{A}^{p,q} : \sigma(\bar{\partial}, dr)\phi = 0 \text{ on } bM\} .$$

Recall how we set up the $\bar{\partial}$-Neumann problem in Chapter 1. If we were to take the operator ϑ as our primary object instead of $\bar{\partial}$, considered as a closed operator on Hilbert space, we would see by the reasoning of Proposition (1.3.2) that $\mathcal{C}^{p,q} = \mathcal{A}^{p,q} \cap \text{Dom}(\vartheta^*)$. We could then formulate the ϑ-Neumann problem in analogy to the $\bar{\partial}$-Neumann problem, and the spaces $\mathcal{C}^{p,q}$ would play the role analogous to $\mathcal{D}^{p,q}$. We wish to derive here some duality relations between these problems.

Recall that the *Hodge star operator* $\star : \mathcal{A}^{p,q} \to \mathcal{A}^{n-p,n-q}$ is defined by the equation $\psi \wedge \star\phi = \langle\psi, \bar{\phi}\rangle \gamma$ where γ is the volume form on M. It is not hard to verify that $\star\star = (-1)^{p+q}$ and $\star\bar{\phi} = \overline{\star\phi}$. Now if $\phi \in \mathcal{A}^{p,q}$ and $\psi \in \mathcal{A}^{p,q+1}$ have compact support, we have

$$(\phi, \vartheta\psi) = (\bar{\partial}\phi, \psi) = \int_M \langle\bar{\partial}\phi, \psi\rangle \gamma = \int_M \bar{\partial}\phi \wedge \star\bar{\psi}$$

$$= \int_M \bar{\partial}(\phi \wedge \star\bar{\psi}) + (-1)^{p+q+1} \int_M \phi \wedge \bar{\partial}\star\bar{\psi} .$$

Now $\int_M \bar\partial(\phi \wedge \star\bar\psi) = \int_M d(\phi \wedge \star\bar\psi) - \int_M \partial(\phi \wedge \star\bar\psi)$; the first term on the right is zero by Stokes' theorem, and the second term is zero because $\phi \wedge \star\bar\psi$ is of type $(n, n-1)$ and hence $\partial(\phi \wedge \star\bar\psi) = 0$. On the other hand,

$$\int_M \phi \wedge \bar\partial\star\bar\psi = (-1)^{p+q} \int_M \phi \wedge \star\,\overline{(\star\partial\star\psi)} = (-1)^{p+q}(\phi, \star\partial\star\psi) .$$

Therefore we have $(\phi, \vartheta\psi) = -(\phi, \star\partial\star\psi)$, and hence:

(5.1.1) PROPOSITION. $\vartheta = -\star\partial\star$.

We now prove the duality of the spaces $\mathcal{C}^{p,q}$ and $\mathcal{D}^{p,q}$.

(5.1.2) PROPOSITION. $\mathcal{C}^{p,q} = \star\,\overline{\mathcal{D}^{n-p,\,n-q}}$.

Proof: $\phi \in \star\,\overline{\mathcal{D}^{n-p,\,n-q}} \Longleftrightarrow \psi = \star\bar\phi \in \mathcal{D}^{n-p,\,n-q} \Longleftrightarrow \sigma(\vartheta, dr)\psi = 0$ on $bM \Longleftrightarrow \sigma(\star\partial\star, dr)\psi = 0$ on $bM \Longleftrightarrow \star\sigma(\bar\partial, dr)\star\bar\psi = 0$ on $bM \Longleftrightarrow \sigma(\bar\partial, dr)\phi = 0$ on $bM \Longleftrightarrow \phi \in \mathcal{C}^{p,q}$. ($\star$ commutes with σ because $\star$ is a zero-order operator and hence is its own symbol.) Q.E.D.

In order to study the relationships on the cohomology level, we show that the spaces $\mathcal{C}^{p,q}$ form a complex under $\bar\partial$.

(5.1.3) PROPOSITION. $\bar\partial\mathcal{C}^{p,q} \subset \mathcal{C}^{p,q+1}$.

Proof: If $\phi \in \mathcal{C}^{p,q}$, ϕ can be written as $\phi = \bar\partial r \wedge \alpha + r\theta$ where $\alpha \in \mathcal{A}^{p,q-1}$, $\theta \in \mathcal{A}^{p,q}$. But then $\bar\partial\phi = \bar\partial r \wedge (\theta - \alpha) + r\bar\partial\theta$, which is in $\mathcal{C}^{p,q+1}$. Q.E.D.

We may therefore form the cohomology

$$H^{(p,q)}(\mathcal{C}) = \{\phi \in \mathcal{C}^{p,q} : \bar\partial\phi = 0\} / \bar\partial\mathcal{C}^{p,q-1} .$$

We also introduce the Dirichlet or zero-boundary-value cohomology,

$$H_0^{(p,q)} = \{\phi \in \mathcal{A}^{p,q} : \bar\partial\phi = 0,\ \phi|bM = 0\} / \bar\partial\{\phi \in \mathcal{A}^{p,q-1} : \phi|bM = 0,\ \bar\partial\phi|bM = 0\} .$$

Moreover, we denote the cohomology of the full $\bar{\partial}$ complex by $H^{(p,q)}(\mathcal{Q})$. (Thus $H^{(p,q)}(\mathcal{Q}) = H^{(p,q)}(\overline{M})$ in the terminology of §4.3, and if M satisfies condition $Z(q)$, $H^{(p,q)}(\mathcal{Q}) \cong \mathcal{H}^{p,q}$.)

(5.1.4) PROPOSITION. $H^{(p,q)}(\mathcal{C}) \cong H_0^{(p,q)}$.

Proof: Suppose $\phi \in \mathcal{Q}^{p,q}$, $\phi|bM = 0$, and $\phi = \bar{\partial}\psi$ with $\psi \in \mathcal{C}^{p,q-1}$. As in Proposition (5.1.3) we write $\psi = \bar{\partial}r \wedge a + r\theta = \bar{\partial}(ra) + r(\bar{\partial}a + \theta)$. Then, setting $\psi_0 = r(\bar{\partial}a + \theta)$, we have $\phi = \bar{\partial}\psi_0$ and $\psi_0|bM = 0$; therefore there is a well-defined injective map of $H_0^{(p,q)}$ into $H^{(p,q)}(\mathcal{C})$. To show surjectivity, suppose $\phi \in \mathcal{C}^{p,q}$ and $\bar{\partial}\phi = 0$; then $\phi = \bar{\partial}(ra) + r(\theta + \bar{\partial}a)$, so ϕ is cohomologous to $r(\theta + \bar{\partial}a)$, which vanishes on bM. Q.E.D.

(5.1.5) PROPOSITION. *If* M *satisfies condition* $Z(q)$, $H^{(p,q)}(\mathcal{Q})$ *is naturally dual to* $H^{(n-p,\,n-q)}(\mathcal{C})$.

Proof: If $\psi \in \mathcal{C}^{n-p,\,n-q}$ and $\bar{\partial}\psi = 0$, we define a linear functional $L_\psi : \{\phi \in \mathcal{Q}^{p,q} : \bar{\partial}\phi = 0\} \to \mathbb{C}$ on cocycles by $L_\psi(\phi) = \int_M \phi \wedge \psi$. If $\phi = \bar{\partial}a$, $a \in \mathcal{Q}^{p,q-1}$, we have

$$L_\psi(\bar{\partial}a) = \int_M \bar{\partial}a \wedge \psi = \int_M \bar{\partial}(a \wedge \psi) + (-1)^{p+q+1} \int_M a \wedge \bar{\partial}\psi \tag{5.1.6}$$

$$= \int_M \bar{\partial}(a \wedge \psi) = \int_{bM} a \wedge \psi - \int_M \partial(a \wedge \psi)$$

by Stokes' theorem. Now $a \wedge \psi \in \mathcal{Q}^{n,\,n-1}$, so $\partial(a \wedge \psi) = 0$. Also, since $\psi|bM = \theta \wedge \bar{\partial}r$ for some θ,

$$\int_{bM} a \wedge \psi = \int_{bM} a \wedge \theta \wedge \bar{\partial}r = -\int_{bM} \bar{\partial}(ra \wedge \theta) + \int_{bM} r\bar{\partial}(a \wedge \theta)$$

$$= -\int_{bM} \bar{\partial}(ra \wedge \theta) = -\int_{bM} d(ra \wedge \theta) + \int_{bM} \partial(ra \wedge \theta)\,.$$

The first term of this last expression is zero by Stokes' theorem, and $\partial(r\alpha \wedge \theta) = 0$ because $\alpha \wedge \theta$ is of type $(n, n-2)$. Therefore $L_\psi(\bar\partial\alpha) = 0$ by (5.1.6), so L_ψ induces a functional on the cohomology group $H^{(p,q)}(\mathcal{A})$. By the same reasoning, $L_\psi = 0$ if $\psi = \bar\partial\beta$. We therefore have a well-defined map $H^{(n-p,n-q)}(\mathcal{C}) \to (H^{(p,q)}(\mathcal{A}))^*$.

To show injectivity, suppose $\psi \in \mathcal{C}^{n-p,n-q}$, $\bar\partial\psi = 0$, and $L_\psi = 0$. Then for all $\bar\partial$-closed ϕ,

$$0 = \int \phi \wedge \psi = (-1)^{p+q} \int \phi \wedge {\star\star}\psi = (-1)^{p+q}(\phi, \star\bar\psi) ,$$

so $\star\bar\psi \perp \mathcal{N}(\bar\partial)$. Since M satisfies condition Z(q), Theorem (3.1.14) implies that $\star\bar\psi = \vartheta\theta$ for some $\theta \in \mathcal{D}^{p,q+1}$. By Proposition (5.1.1), $\psi = -\bar\partial\star\bar\theta$, and by Proposition (5.1.2), $\star\bar\theta \in \mathcal{C}^{n-p,n-q-1}$. Thus the image of ψ in $H^{(n-p,n-q)}(\mathcal{C})$ is zero. To show surjectivity, suppose $\phi \in \mathcal{A}^{p,q}$, $\bar\partial\phi = 0$, and $L_\psi(\phi) = (\phi, \star\bar\psi) = 0$ for all $\psi \in \mathcal{C}^{n-p,n-q}$ with $\bar\partial\psi = 0$. Again, by Propositions (5.1.1) and (5.2.2), $\vartheta\star\bar\psi = 0$ and $\star\bar\psi \in \mathcal{D}^{p,q}$, so $\phi \perp \ker(\bar\partial^*)$. By Theorem (3.1.14), $\phi = \bar\partial\theta$ for some $\theta \in \mathcal{A}^{p,q-1}$, so ϕ is cohomologous to zero. Q.E.D.

Combining Propositions (5.1.4) and (5.1.5), we obtain the following result, which is a first cousin of the Serre duality theorem [40]:

(5.1.7) PROPOSITION. *If* M *satisfies condition* Z(q), *then* $H_0^{(n-p,n-q)} \cong (H^{(p,q)}(\mathcal{A}))^*$.

The results of this section, in particular Proposition (5.1.2), show that the star operator yields an isomorphism between the $\bar\partial$-Neumann problem for (p,q)-forms and the ϑ-Neumann problem for $(n-p, n-q)$-forms. Suppose M satisfies conditions Z(q) and Z(n−q), so these problems are both solvable for (p,q)-forms. Then $\bar\partial^*\bar\partial$ and $\vartheta^*\vartheta$ both have closed range. Since $\bar\partial\vartheta^* = \bar\partial^2 = 0$, these ranges are orthogonal, and it follows without difficulty that $T = \bar\partial^*\bar\partial + \vartheta^*\vartheta$ also has closed range.

The operator T was considered by Garabedian and Spencer [12]; it follows from the argument of Gaffney [11] used to prove Proposition (1.3.8) that T is a self-adjoint extension of $\square$. But the nullspace of T consists of all $\phi \in H_0^{p,q}$ such that $\bar{\partial}\phi = \vartheta\phi = 0$ (with no boundary conditions!), and this space is always infinite-dimensional. The fact that we can nonetheless show that Range (T) is closed is a curious example of "existence without estimates."

2. *The induced boundary complex*

We now introduce spaces $\mathcal{B}^{p,q}$ of forms on bM according to the following three equivalent definitions:

(1) $\mathcal{B}^{p,q}$ is the space of (smooth) sections of the vector bundle $\Lambda^{p,q}CT^*M \cap \Lambda^{p+q}CT^*bM$ on bM.

(2) $\mathcal{B}^{p,q}$ is the space of (p,q)-forms restricted to bM which are pointwise orthogonal to the ideal generated by $\bar{\partial}r$ (i.e., to all forms of the type $\bar{\partial}r \wedge \theta$).

(3) $\mathcal{B}^{p,q}$ is the space of restrictions of elements of $\mathcal{D}^{p,q}$ to bM.

The equivalence of (1) and (2) is easy to check, and the equivalence of (2) and (3) is also clear by Lemma (2.3.2). (1) says that $\mathcal{B}^{p,q}$ is the space of tangential (p,q)-forms on bM.

Using the language of sheaves, we may express $\mathcal{B}^{p,q}$ in yet another way, which is clearly equivalent to (2):

(4) Let $\tilde{\mathcal{A}}^{p,q}$ and $\tilde{\mathcal{C}}^{p,q}$ denote the sheaves of germs of $\mathcal{A}^{p,q}$ and $\mathcal{C}^{p,q}$ on $\overline{M}$, respectively. Then there is a natural injection $0 \to \tilde{\mathcal{C}}^{p,q} \to \tilde{\mathcal{A}}^{p,q}$. The quotient sheaf $\tilde{\mathcal{B}}^{p,q} = \tilde{\mathcal{A}}^{p,q}/\tilde{\mathcal{C}}^{p,q}$ is a locally free sheaf supported on bM, and $\mathcal{B}^{p,q}$ is its space of sections.

In view of Proposition (5.1.3), we have the following commutative diagram:

$$\begin{array}{ccccccccc}
0 & \longrightarrow & \tilde{\mathcal{C}}^{p,q+1} & \longrightarrow & \tilde{\mathcal{A}}^{p,q+1} & \longrightarrow & \tilde{\mathcal{B}}^{p,q+1} & \longrightarrow & 0 \\
 & & \bar{\partial}\uparrow & & \bar{\partial}\uparrow & & & & \\
0 & \longrightarrow & \tilde{\mathcal{C}}^{p,q} & \longrightarrow & \tilde{\mathcal{A}}^{p,q} & \longrightarrow & \tilde{\mathcal{B}}^{p,q} & \longrightarrow & 0 .
\end{array}$$

There is therefore induced a quotient map $\tilde{\mathcal{B}}^{p,q} \to \tilde{\mathcal{B}}^{p,q+1}$ which we denote by $\bar{\partial}_b$. $\bar{\partial}_b$ may be explicitly described on sections as follows: if $\phi \in \mathcal{B}^{p,q}$, choose $\phi' \in \mathcal{D}^{p,q}$ such that $\phi'|bM = \phi$. Then $\bar{\partial}_b\phi$ is the orthogonal projection of $\bar{\partial}\phi'|bM$ onto $\mathcal{B}^{p,q}$. It is easy to check that this is independent of the choice of ϕ'.

Since $\bar{\partial}^2 = 0$, it follows that $\bar{\partial}_b^2 = 0$, so we have the *boundary complex*

$$(5.2.1) \qquad 0 \longrightarrow \mathcal{B}^{p,0} \xrightarrow{\bar{\partial}_b} \mathcal{B}^{p,1} \xrightarrow{\bar{\partial}_b} \cdots \xrightarrow{\bar{\partial}_b} \mathcal{B}^{p,n-1} \longrightarrow 0 .$$

(Note that $\mathcal{B}^{p,n} = 0$.) We denote the cohomology of this complex by $H^{(p,q)}(\mathcal{B})$.

The operator $\bar{\partial}_b$ was studied by H. Lewy [30] in the case $n = 2$ in connection with the problem of finding holomorphic extensions of smooth functions on real submanifolds of $\mathbf{C}^2$, a problem which we shall consider in the next section. This work was extended by Kohn and Rossi [28], who formalized the notion of the boundary complex, and to whom most of the results in the first three sections of this chapter are due. More recently there has developed a general theory of boundary complexes associated to Neumann problems for overdetermined systems, cf. Sweeney [43]. The philosophy behind this is to reduce questions about boundary value problems on M to the study of operators on bM. which is a compact manifold without boundary. Much progress in this area has recently been made by Kuranishi. For the present, however, we content ourselves with the following two propositions.

(5.2.2) PROPOSITION. *If* $\phi \in \mathcal{B}^{p,q}$, *then* $\bar{\partial}_b\phi = 0$ *if and only if* $\vartheta \star \bar{\phi}' \in \mathcal{D}^{n-p,n-q-1}$ *for any smooth extension* ϕ' *of* ϕ.

Proof: $\bar{\partial}_b\phi = 0$ means that $\bar{\partial}\phi' \in \mathcal{C}^{p,q+1}$; by Proposition (5.1.2), this happens if and only if $\vartheta \star \phi' \in \mathcal{D}^{n-p,n-q-1}$. Q.E.D.

(5.2.3) PROPOSITION. *If* M *satisfies conditions* Z(q) *and* Z(n−q−1), *then* $H^{(p,q)}(\mathcal{B})$ *is finite-dimensional, and the range of* $\bar{\partial}_b : \mathcal{B}^{p,q-1} \to \mathcal{B}^{p,q}$ *is closed in the* $\mathcal{C}^\infty$ *topology. (It will also follow that the range of the Hilbert space operator* $\bar{\partial}_b$ *is closed once we have proved the regularity theorems of* §5.4.)

Proof: Condition Z(q) implies that $\dim H^{(p,q)}(\mathcal{Q}) < \infty$, and by Proposition (5.1.5), condition Z(n−q−1) implies $\dim H^{(p,q+1)}(\mathcal{C}) < \infty$. But the exact sequence of groups $0 \to \mathcal{C}^{p,q} \to \mathcal{Q}^{p,q} \to \mathcal{B}^{p,q} \to 0$ (where the third arrow is restriction followed by projection) induces the exact cohomology sequence $H^{(p,q)}(\mathcal{Q}) \to H^{(p,q)}(\mathcal{B}) \to H^{(p,q+1)}(\mathcal{C})$ by the usual diagram chase, which shows that $\dim H^{(p,q)}(\mathcal{B}) \le \dim H^{(p,q)}(\mathcal{Q}) + \dim H^{(p,q+1)}(\mathcal{C}) < \infty$. Next, in $\mathcal{B}^{p,q}$ we can write $\mathcal{N}(\bar{\partial}_b) = \text{Range}\,(\bar{\partial}_b) \oplus K$ where $\dim K < \infty$; then, setting $B = \mathcal{B}^{p,q-1}/\mathcal{N}(\bar{\partial}_b)$, $\bar{\partial}_b \oplus I : B \oplus K \to \mathcal{N}(\bar{\partial}_b)$ is a continuous bijection of Fréchet spaces. By the closed graph theorem, it is an isomorphism, so $\text{Range}\,(\bar{\partial}_b) = \bar{\partial}_b \oplus I(B \oplus \{0\})$ is closed. Q.E.D.

Note what conditions Z(q) and Z(n−q−1) mean together: max (q+1, n−q) eigenvalues of the Levi form have the same sign, or there are min (q+1, n−q) pairs of eigenvalues with opposite signs. This symmetry with respect to signs is indicative of the general fact that, crudely speaking, the boundary complex doesn't know whether bM is being considered as the boundary of M or the boundary of the complement of M. Notice also that for $n = 2$ and $q = 1$ these conditions are never satisfied. It is this situation that led H. Lewy to his famous example of a smooth differential equation without solution [30a], which has had such great influence on the research of the last fifteen years.

3. $\bar{\partial}$-*closed extensions of forms*

In this section we consider the following problem: given a (p,q)-form ϕ on bM (i.e., ϕ is the restriction to bM of an element of $\mathcal{Q}^{p,q}$), does there exist $\psi \in \mathcal{Q}^{p,q}$ with $\psi|bM = \phi$ and $\bar{\partial}\psi = 0$? Clearly a necessary condition is that $\bar{\partial}_b\phi_0 = 0$ where ϕ_0 is the projection of ϕ on

$\mathcal{B}^{p,q}$, for we can write $\psi = \psi_0 + \psi_1$ where $\psi_0 \in \mathcal{D}^{p,q}$ and $\psi_1 \in \mathcal{C}^{p,q}$; then $\psi_0|bM = \phi_0$ and $\bar{\partial}\psi_0 = -\bar{\partial}\psi_1 \in \mathcal{C}^{p,q+1}$ by Proposition (5.1.3). On the other hand, our machinery gives us no leverage on the components of ϕ and ψ which contain $\bar{\partial}r$. We therefore propose the following weaker problem: Given a (p,q)-form ϕ on bM, does there exist $\psi \in \mathcal{A}^{p,q}$ with $\bar{\partial}\psi = 0$ such that $\psi|bM$ and ϕ have the same projection on $\mathcal{B}^{p,q}$? If so, we call ψ a *weak* $\bar{\partial}$*-closed extension* of ϕ. Clearly for this problem it suffices to consider $\phi \in \mathcal{B}^{p,q}$, and we can give necessary and sufficient conditions for its solution under a pseudoconvexity hypothesis.

(5.3.1) THEOREM. *If* M *satisfies condition* Z(n−q−1), *there is a weak* $\bar{\partial}$*-closed extension of* $\phi \in \mathcal{B}^{p,q}$ *if and only if* $\bar{\partial}_b\phi = 0$ and $\int_{bM}\theta\wedge\phi = 0$ *for all* $\theta \in \mathcal{H}^{n-p,n-q-1}$.

Proof: Suppose $\bar{\partial}_b\phi = 0$ and $\int_{bM}\theta\wedge\phi = 0$ for all $\theta \in \mathcal{H}^{n-p,n-q-1}$; let $\phi' \in \mathcal{D}^{p,q}$ extend ϕ. Then if $\theta \in \mathcal{H}^{n-p,n-q-1}$,

$$(\theta, \vartheta\star\bar{\phi}') = (-1)^{p+q}(\theta, \overline{\star\bar{\partial}\phi'}) = \int_M \theta\wedge\bar{\partial}\phi' = (-1)^{p+q+1}\int_M \bar{\partial}(\theta\wedge\phi')$$

since $\bar{\partial}\theta = 0$, and since $\theta\wedge\phi'$ is of type (n, n−1),

$$\int_M \bar{\partial}(\theta\wedge\phi') = \int_M d(\theta\wedge\phi') = \int_{bM} \theta\wedge\phi = 0 .$$

Thus $\vartheta\star\bar{\phi}' \perp \mathcal{H}^{n-p,n-q-1}$, and by Proposition (5.2.2), $\vartheta\star\bar{\phi}' \in \mathcal{D}^{n-p,n-q-1}$, so $\bar{\partial}^*\vartheta\star\bar{\phi}' = \vartheta^2\star\bar{\phi}' = 0$. Hence by the analogue of Proposition (3.1.15) for $\bar{\partial}^*$, there exists $\alpha \in \mathcal{D}^{n-p,n-q}$ with $\vartheta\alpha = \vartheta\star\bar{\phi}'$. Set $\psi = \phi' + (-1)^{p+q}\star\bar{\alpha}$. By Proposition (5.1.2), $\star\bar{\alpha} \in \mathcal{C}^{p,q}$, so the projection of ψ on $\mathcal{B}^{p,q}$ is ϕ. Also,

$$\bar{\partial}\psi = \bar{\partial}\phi' + (-1)^{p+q}\bar{\partial}\star\bar{\alpha} = \bar{\partial}\phi' + \star\overline{\vartheta\alpha} = \bar{\partial}\phi' + \overline{\star\vartheta\star\bar{\phi}'} = \bar{\partial}\phi' - \bar{\partial}\phi' = 0 .$$

Thus the problem is solved.

Conversely, suppose such a ψ exists; we may write $\psi|bM = \phi + a \wedge \bar\partial r$ for some a. Since $\bar\partial\psi = 0$, we have $\vartheta \star \bar\psi = 0$, so by the same reasoning as above, for $\theta \in \mathcal{H}^{n-p,n-q-1}$, $0 = (\theta, \vartheta\star\bar\psi) = \int_{bM} \theta \wedge \phi + \int_{bM} \theta \wedge a \wedge \bar\partial r$. The second term is zero by the argument in the proof of Proposition (5.1.5), so $\int_{bM} \theta \wedge \phi = 0$. Finally, if ϕ' extends ϕ, then $\phi' - \psi \in \mathcal{C}^{p,q}$, so by Proposition (5.1.3), $\bar\partial\phi' = \bar\partial(\phi' - \psi) \in \mathcal{C}^{p,q+1}$, which shows that $\bar\partial_b \phi = 0$. Q.E.D.

Remark. If $M = \{z \in \mathbf{C}^n : |z| < 1\}$ – in which case, for $q > 0$, condition $Z(q)$ is satisfied and $\mathcal{H}^{p,q} = 0$ – it follows from the work of Folland [9] that the strong extension problem (that is, find ψ with $\bar\partial\psi = 0$ and $\psi|bM = \phi$) is solvable for all ϕ in a dense subspace of $\mathfrak{N}(\bar\partial_b) \subset \mathcal{B}^{p,q}$ $(q < n-1)$. However, even in this simple case, convergence questions make it appear unlikely that the strong extension problem is solvable in general.

In the case of functions we can say more. First we note that $\mathcal{B}^{0,0}$ is the space of all smooth functions on bM, so the weak and strong extension problems coincide.

(5.3.2) THEOREM. *Suppose that* M *is connected and satisfies condition* $Z(n-1)$, *and that there exists a non-constant holomorphic function* h *on* M *which is smooth up to the boundary. Then every* $f \in \mathcal{B}^{0,0}$ *with* $\bar\partial_b f = 0$ *has a holomorphic extension to* M.

Proof: Given such an f, we have $\bar\partial_b(f^k) = 0$ for all $k \geq 0$. Let $\lambda = \dim \mathcal{H}^{n,n-1} < \infty$, let $\theta_1, \ldots, \theta_\lambda$ be a basis for $\mathcal{H}^{n,n-1}$, and let R denote the operation of restriction to bM. For $0 \leq j \leq \lambda$; $1 \leq i \leq \lambda$; set $c_{ij} = \int_{bM} f^j R\theta_i$, and choose complex numbers $a_0, a_1, \ldots, a_\lambda$, such that $\Sigma_j a_j c_{ij} = 0$ for all i; we may assume $a_\lambda = 1$. Set $P(X) = \Sigma_0^\lambda a_j X^j$. Then $g_0 = P(f)$ has the properties $\bar\partial_b g_0 = 0$ and $\int_{bM} g_0 R\theta = 0$ for all $\theta \in \mathcal{H}^{n,n-1}$, so by Theorem (5.3.1), g_0 has a holomorphic extension to M which we denote by g. Next, if h is our non-constant holomorphic function, we have $\bar{\partial}_b((Rh)^j f) = 0$ for $j \geq 0$, so by the same procedure we may choose numbers $b_0, \ldots, b_\lambda$ such that $(\Sigma b_j (Rh)^j)f$ has a holomorphic extension F to M. If we set $G = \Sigma b_j h^j$, then $(RG)f = RF$, so $R(G^\lambda P(F/G)) = R(G^\lambda) P(RF/RG) = R(G^\lambda)Rg = R(G^\lambda g)$. Thus $G^\lambda P(F/G)$ and $G^\lambda g$ have the same boundary values, and they are both holomorphic; therefore they are equal. Hence $P(F/G) = g$ on $M - \{x : G(x) = 0\}$. But since h is non-constant and M is connected, h assumes infinitely many values, so codim $\{x : G(x) = 0\} \geq 1$. Since P is a monic polynomial with constant coefficients, this implies F/G is locally bounded on M and hence holomorphic. F/G is thus the desired extension of f. Q.E.D.

(5.3.3) COROLLARY. *Under the hypotheses of Theorem* (5.3.2), bM *is connected.*

Proof: Let $f \in \mathcal{B}^{0,0}$ be locally constant. Then $\bar{\partial}_b f = 0$, so f has a holomorphic extension F to M which must be locally constant. Since M is connected, F is constant. Thus f must be constant, so bM is connected. Q.E.D.

(5.3.4) COROLLARY. *Suppose* M *is connected and satisfies condition* $Z(n-1)$. *If there exists* $g \in \mathcal{B}^{0,0}$ *which is not locally constant but* $\bar{\partial}_b g = 0$, *then every* $f \in \mathcal{B}^{0,0}$ *with* $\bar{\partial}_b f = 0$ *has a holomorphic extension to* M.

Proof: As in the proof of Theorem (5.3.2), there is a polynomial P such that $P(g)$ has a holomorphic extension h to M. By the hypothesis on on g, h is non-constant, so we can apply Theorem (5.3.2). Q.E.D.

(5.3.5) COROLLARY. *If* M *is connected and satisfies condition* $Z(n-1)$ *and* bM *is connected, then every* $f \in \mathcal{B}^{0,0}$ *with* $\bar{\partial}_b f = 0$ *has a holomorphic extension to* M.

Proof: If every such f is constant, the conclusion is trivial. If not, we can apply Corollary (5.3.4). Q.E.D.

(5.3.6) COROLLARY. *If* M *is connected and satisfies condition* Z(n−1) *but* bM *is disconnected, then every holomorphic function on* M *is constant and every* $f \in \mathcal{B}^{0,0}$ *with* $\bar\partial_b f = 0$ *is locally constant.*

Proof: Corollary (5.3.3) implies that every holomorphic function on M which is smooth up to the boundary is constant; the shrinking argument in Corollary (3.2.26) then proves the result for arbitrary holomorphic functions. The second assertion follows from Corollary (5.3.3) and the proof of Corollary (5.3.4). Q.E.D.

For a different formulation of Theorem (5.3.1) and a theorem on meromorphic extensions of sections of holomorphic vector bundles, see Kohn and Rossi [28]. Other results along these lines, including theorems relating the boundary cohomology to the smooth cohomology of the interior and exterior manifolds M and $M' - \bar M$, have recently been obtained by A. Andreotti and C. D. Hill; cf. [2] and [3]. Further extension theorems may be found in the work of S. Greenfield [13b] and R. Nirenberg [36b].

4. *The abstract model*

We can form the formal adjoint $\vartheta_b : \mathcal{B}^{p,q} \to \mathcal{B}^{p,q-1}$ of $\bar\partial_b$ – which, since bM is a compact manifold without boundary, can be identified with its Hilbert space adjoint – and the associated Laplacian $\Box_b = \bar\partial_b \vartheta_b + \vartheta_b \bar\partial_b$. $\Box_b$ is not elliptic, a fact which reflects the non-coerciveness of the $\bar\partial$-Neumann problem: one easily verifies that the real cotangent vector $i(\bar\partial - \partial)r$ is characteristic at each point. However, Kohn [23] showed that under suitable pseudoconvexity conditions, the $\bar\partial_b$ complex can be studied by potential-theoretic methods. We proceed to investigate this problem.

In this section we shall be studying the $\bar{\partial}_b$ complex for its own sake, so we begin by reformulating it in a way independent of the interior manifold M. Let X be a compact, orientable, real manifold of dimension 2n−1. A *partially complex structure* on X is an (n−1)-dimensional sub-bundle S of CTX such that (1) $S \cap \bar{S} = \{0\}$, (2) if L, L' are local sections of S then so is $[L, L']$. (In particular, if $X = bM$ where M is a complex manifold, then $S = (\Pi_{1,0}CTM) \cap (CTX)$ defines a partially complex structure on X.) If X is partially complex, we define the vector bundle $B^{p,q}$ $(0 \le p, q \le n-1)$ by $B^{p,q} = \Lambda^p S^* \otimes \Lambda^q \bar{S}^*$, which we can identify with a sub-bundle of $\Lambda^{p+q} CT^*X$. We denote by $\mathcal{B}^{p,q}$ the space of smooth sections of $B^{p,q}$, and we define $\bar{\partial}_b : \mathcal{B}^{p,q} \to \mathcal{B}^{p,q+1}$ as follows. If $\phi \in \mathcal{B}^{p,0}$, $\bar{\partial}_b \phi$ is defined by

$$<\bar{\partial}_b \phi, (L_1 \wedge \ldots \wedge L_p) \otimes V> = V<\phi, L_1 \wedge \ldots \wedge L_p>$$

for all sections $L_1, \ldots, L_p$ of S and V of $\bar{S}$. We then extend $\bar{\partial}_b$ to $\mathcal{B}^{p,q}$, $q > 0$, as a derivation. In other words, if $\phi \in \mathcal{B}^{p,q}$,

$$(q+1)<\bar{\partial}_b \phi, (L_1 \wedge \ldots \wedge L_p) \otimes (V_1, \ldots, V_{q+1})>$$

$$= \sum_{j=1}^{q+1} (-1)^{j+1} V_j <\phi, (L_1, \ldots, L_p) \otimes (V_1 \wedge \ldots \hat{V}_j \ldots \wedge V_{q+1})>$$

$$+ \sum_{i<j} (-1)^{i+j} <\phi, (L_1 \wedge \ldots \wedge L_p) \otimes ([V_i, V_j] \wedge V_1 \wedge \ldots \hat{V}_i \ldots \hat{V}_j \ldots \wedge V_{q+1})> .$$

Let $L_1, \ldots, L_{n-1}$ be a local basis for sections of S over $U \subset X$, so $\bar{L}_1, \ldots, \bar{L}_{n-1}$ is a local basis for sections of $\bar{S}$. Since $S \oplus \bar{S}$ has (complex) codimension one in CTX, we may choose a local section N of CTX such that $L_1, \ldots, L_{n-1}, \bar{L}_1, \ldots, \bar{L}_{n-1}$, N span CTX, and we may assume that N is purely imaginary. Then the matrix (c_{ij}) defined by

$$[L_i, \bar{L}_j] = \sum a_{ij}^k L_k + \sum b_{ij}^k \bar{L}_k + c_{ij} N$$

is Hermitian; it is called the *Levi form*. It is, of course, highly non-invariant; however, its essential features are invariant.

(5.4.1) PROPOSITION. *The number of non-zero eigenvalues and the absolute value of the signature of* (c_{ij}) *at each point are independent of the choice of* $L_1, \ldots, L_{n-1}, N$.

Proof: If η is a purely imaginary non-vanishing one-form on U which annihilates $S \oplus \bar{S}$, it is easily verified that $c_{ij} = 2\langle d\eta, L_i \wedge \bar{L}_j\rangle / \langle \eta, N\rangle$. But the Hermitian form $\langle d\eta, L \wedge \bar{L}'\rangle$ on S is invariantly defined once η is given, and $\langle \eta, N\rangle$ is a nonvanishing real function, which does not affect the number of non-zero eigenvalues or the absolute value of the signature. Q.E.D.

In view of Proposition (5.4.1), it makes sense to require that the Levi form have $\max(q+1, n-q)$ eigenvalues of the same sign or $\min(q+1, n-q)$ pairs of eigenvalues with opposite signs at each point. If this is true, we say that X satisfies *condition* $Y(q)$. (Note that if $X = bM$, then the new and old Levi forms coincide up to sign and normalization, and X satisfies condition $Y(q)$ if and only if M satisfies conditions $Z(q)$ and $Z(n-q-1)$.)

N may in fact be defined globally, for if we restrict our coordinate transformations to those which preserve S and the orientation, a local choice of $+$ or $-$ in the N direction extends to a global one. Let us choose a Hermitian metric on $\mathbb{C}TX$ such that $S, \bar{S}$, and N are mutually orthogonal. We may then assume that $L_1, \ldots, L_{n-1}, \bar{L}_1, \ldots, \bar{L}_{n-1}, N$ are orthonormal. We can define the Sobolev spaces $H_s^{p,q}$ for all real s by completing $\mathcal{B}^{p,q}$ appropriately, cf. Appendix, §2, and define the adjoint operator ϑ_b and the Laplacian $\Box_b = \bar\partial_b \vartheta_b + \vartheta_b \bar\partial_b$. If $\omega_1, \ldots, \omega_{n-1}, \bar\omega_1, \ldots, \bar\omega_{n-1}, \eta$ is the dual basis to $L_1, \ldots, L_{n-1}, \bar{L}_1, \ldots, \bar{L}_{n-1}, N$, we write $\phi \in \mathcal{B}^{p,q}$ as $\phi = \Sigma_{IJ} \phi_{IJ} \omega^I \wedge \bar\omega^J$ in the usual formalism, and we have the formulas for $\bar\partial_b$ and ϑ_b:

$$\bar\partial_b \phi = (-1)^p \sum_{kIJK} \epsilon^K_{kJ} \bar{L}_k(\phi_{IJ}) \omega^I \wedge \bar\omega^K + \text{terms of order zero},$$

$$\vartheta_b \phi = (-1)^{p+1} \sum_{kIHJ} \epsilon^J_{kH} L_k(\phi_{IJ}) \omega^I \wedge \bar\omega^H + \text{terms of order zero}.$$

By analogy with the $\bar{\partial}$-Neumann problem, we define the Hermitian form Q_b on $\mathcal{B}^{p,q}$ by

$$Q_b(\phi,\psi) = (\bar{\partial}_b\phi, \bar{\partial}_b\psi) + (\vartheta_b\phi, \vartheta_b\psi) + (\phi,\psi) = ((\Box_b + I)\phi, \psi) .$$

Assuming X satisfies condition Y(q), we shall derive some *a priori* estimates for Q_b. The proof of these estimates is very similar to that of Theorem (3.2.10), so we only sketch the details.

To save space, we abbreviate $\Sigma_{kIJ} \|L_k\phi_{IJ}\|^2 + \|\phi\|^2$ by $\|\phi\|_L^2$ and $\Sigma_{kIJ} \|\bar{L}_k\phi_{IJ}\|^2 + \|\phi\|^2$ by $\|\phi\|_{\bar{L}}^2$.

(5.4.2) LEMMA. *Let* $f \in \mathcal{B}^{0,0}$ *be fixed. If the Levi form at* $x_0 \in X$ *has one nonzero eigenvalue, there is a neighborhood* V *of* x_0 *such that*

$$|\mathrm{Re}\,(Nu, fu)| \lesssim (\sup_V |f|)\,(\|u\|_L^2 + \|u\|_{\bar{L}}^2) + \|u\|^2$$

uniformly for $u \in \mathcal{B}^{0,0}$ *with support in* V.

Proof: Choose the L_i's so that the Levi form is diagonal at x_0 with non-zero eigenvalue $c_{11}(x_0)$, and choose V so that $|c_{11}| \geq C^{-1} > 0$ on V. Then if u is supported in V,

$$Nu = \frac{1}{c_{11}} [L_1, \bar{L}_1] u + \mathcal{O}(\|u\|_L + \|u\|_{\bar{L}}) ,$$

so

$$|\mathrm{Re}\,(Nu,fu)| \leq \left|\left(\frac{1}{c_{11}} L_1\bar{L}_1u, fu\right)\right| + \left|\left(\frac{1}{c_{11}} \bar{L}_1L_1u, fu\right)\right| + \mathcal{O}((\|u\|_L + \|u\|_{\bar{L}})\|u\|)$$
$$\leq \left|\left(\frac{1}{c_{11}} \bar{L}_1u, f\bar{L}_1u\right)\right| + \left|\left(\frac{1}{c_{11}} L_1u, fL_1u\right)\right| + \mathcal{O}((\|u\|_L + \|u\|_{\bar{L}})\|u\|)$$
$$\leq C(\sup_V |f|)(\|u\|_L^2 + \|u\|_{\bar{L}}^2) + (sc)(\|u\|_L^2 + \|u\|_{\bar{L}}^2) + (\ell c)\|u\|^2 ,$$

where we may choose $(sc) \leq \sup_V |f|$. Q.E.D.

(5.4.3) LEMMA. *Under the hypothesis of Lemma* (5.4.2), *there is a neighborhood* V *of* x_0 *such that*

$$\|u\|_L^2 \lesssim \|u\|_{\bar L}^2 + |\mathrm{Re}\,(Nu,u)| + \|u\|^2 ,$$

uniformly for all $u \in \mathcal{B}^{0,0}$ *supported in* V.

Proof: Choose the L_i's so the Levi form is diagonal at x_0, i.e., $c_{ij} = \lambda_i\delta_{ij} + b_{ij}$ where $b_{ij}(x_0) = 0$. Then we have

$$\begin{aligned}\|L_iu\|^2 &= (L_iu,L_iu) = -\mathrm{Re}\,(\bar L_iL_iu,u) + \mathcal{O}(\|L_iu\|\,\|u\| + \|u\|^2)\\ &= -\mathrm{Re}\,(L_i\bar L_iu,u) + \mathrm{Re}\,([\bar L_i,L_i]u,u) + \mathcal{O}(\|L_iu\|\,\|u\| + \|u\|^2)\\ &= (\bar L_iu,\bar L_iu) + \lambda_i\,\mathrm{Re}\,(Nu,u) + \mathrm{Re}\,(b_{ii}Nu,u) + \mathcal{O}(\|L_iu\|\,\|u\| + \|\bar L_iu\|\,\|u\| + \|u\|^2)\\ &\lesssim \|\bar L_iu\|^2 + |\lambda_i|\,|\mathrm{Re}\,(Nu,u)| + |\mathrm{Re}\,(Nu,\overline{b_{ii}}u)| + (sc)\|L_iu\|^2 + (\ell c)\|u\|^2 .\end{aligned}$$

Since $b_{ii}(x_0) = 0$, by Lemma (5.4.2) we can choose V small enough that $|\mathrm{Re}\,(Nu,\overline{b_{ii}}u)| \leq (sc)(\|u\|_L^2 + \|u\|_{\bar L}^2) + (\ell c)\|u\|^2$. Summing on i, we are done. Q.E.D.

(5.4.4) LEMMA. *Under the hypothesis of Lemma* (5.4.2), *given* $\delta > 0$ *there exists a neighborhood* V *of* x_0 *such that*

$$\|\bar L_ku\|^2 \geq -\lambda_k\,\mathrm{Re}\,(Nu,u) - \delta\,(\|u\|_L^2 + \|u\|_{\bar L}^2) - (\ell c)\|u\|^2 .$$

uniformly for $u \in \mathcal{B}^{0,0}$ *with support in* V.

Proof: Repeat the calculation in the proof of Lemma (5.4.3), substituting $\bar L_k$ for L_i and omitting the last inequality. Q.E.D.

(5.4.5) LEMMA. *Under the hypothesis of Lemma* (5.4.2), *given* $\delta > 0$ *there exists a neighborhood* V *of* x_0 *such that*

$$Q_b(\phi,\phi) = \|\phi\|_{\bar L}^2 + \sum_{IJ}\sum_{k\in J}\lambda_k\,\mathrm{Re}\,(N\phi_{IJ},\phi_{IJ}) + R(\phi) + \mathcal{O}(\|\phi\|_{\bar L}\,\|\phi\| + \|\phi\|^2) .$$

for all $\phi \in \mathcal{B}^{p,q}$ *with support in* V, *where* $|R(\phi)| \le \delta(\|\phi\|_L^2 + \|\phi\|_{\bar{L}}^2)$ and $\lambda_1, \ldots, \lambda_{n-1}$ *are the eigenvalues of the Levi form at* x_0.

Proof: The proof is almost identical to that of Lemma (3.2.3). The final steps are accomplished with the aid of Lemma (5.4.2), which replaces the boundary integrals. Details are left to the reader. Q.E.D.

(5.4.6) THEOREM. *Suppose X satisfies condition* Y(q). *For any* $x_0 \in X$, *there is a neighborhood* V *of* x_0 *such that*

$$\|\phi\|_L^2 + \|\phi\|_{\bar{L}}^2 + \sum_{IJ} |\mathrm{Re}\,(N\phi_{IJ}, \phi_{IJ})| \lesssim Q_b(\phi,\phi)$$

uniformly for $\phi \in \mathcal{B}^{p,q}$ *with support in* V.

(Note: we shall not make use of the term $\Sigma|\mathrm{Re}\,(N\phi_{IJ}, \phi_{IJ})|$, but it is necessary for the proof. We could also prove the converse by an argument similar to that in Theorem (3.2.21).)

Proof: First note that condition Y(q) always implies the existence of a non-zero eigenvalue, so we may choose V so that the conclusions of Lemmas (5.1.3, 4, 5) hold for suitably small δ. As usual, we choose $L_1, \ldots, L_{n-1}$ so the Levi form at x_0 is diagonal with eigenvalues $\lambda_1, \ldots, \lambda_{n-1}$.

For any two disjoint subsets $\mathcal{P}$ and $\mathcal{N}$ of the set $\{(I,J)\}$ of pairs of multi-indices, we define

$$\mathcal{B}^{p,q}(\mathcal{P},\mathcal{N}) = \{\phi \in \mathcal{B}^{p,q} \text{ with support in } V\colon \mathrm{Re}\,(N\phi_{IJ}, \phi_{IJ}) > 0 \text{ when } (I,J) \in \mathcal{P},\ \mathrm{Re}\,(N\phi_{IJ}, \phi_{IJ}) < 0 \text{ when } (I,J) \in \mathcal{N}, \text{ and } \mathrm{Re}\,(N\phi_{IJ}, \phi_{IJ}) = 0 \text{ when } (I,J) \notin \mathcal{P} \cup \mathcal{N}.\}$$

Since $\{\phi \in \mathcal{B}^{p,q}\colon \operatorname{supp}\phi \subset V\} = \bigcup_{\mathcal{P},\mathcal{N}} \mathcal{B}^{p,q}(\mathcal{P},\mathcal{N})$ and this union is finite, it suffices to prove the theorem for $\phi \in \mathcal{B}^{p,q}(\mathcal{P},\mathcal{N})$ for fixed $\mathcal{P},\mathcal{N}$.

Define $\mathcal{O} = \{(I,J,k) : \lambda_k < 0$ when $(I,J) \in \mathcal{P}$ and $\lambda_k > 0$ when $(I,J) \in \mathcal{N}\}$. Given ϵ, $0 < \epsilon < 1$, by Lemma (5.4.4) we obtain

$$\begin{aligned} \|\phi\|_{\bar{L}}^2 &\geq \epsilon \|\phi\|_{\bar{L}}^2 + (1-\epsilon) \sum_{(IJk)\in\mathcal{O}} \|\bar{L}_k \phi_{IJ}\|^2 \\ &\geq \epsilon \|\phi\|_{\bar{L}}^2 - (1-\epsilon) \sum_{(IJk)\in\mathcal{O}} \lambda_k \operatorname{Re}(N\phi_{IJ}, \phi_{IJ}) \\ &\qquad - \delta(\|\phi\|_L^2 + \|\phi\|_{\bar{L}}^2) - (\ell c)\|\phi\|^2 . \end{aligned}$$

Substituting this into the formula of Lemma (3.2.5),

$$\begin{aligned} Q_b(\phi,\phi) \geq \epsilon \|\phi\|_{\bar{L}}^2 + \sum_{IJ} a_{IJ} \operatorname{Re}(N\phi_{IJ}, \phi_{IJ}) - 2\delta(\|\phi\|_L^2 + \|\phi\|_{\bar{L}}^2) \\ - \mathcal{O}(\|\phi\|_{\bar{L}} \|\phi\| + \|\phi\|^2) \end{aligned}$$

where

$$a_{IJ} = \sum_{k\in J,(IJk)\notin\mathcal{O}} \lambda_k - (1-\epsilon) \sum_{k\notin J,(IJk)\in\mathcal{O}} \lambda_k + \epsilon \sum_{k\in J,(IJk)\in\mathcal{O}} \lambda_k .$$

Now condition $Y(q)$ implies that for each J, either (1) there exist $j,k \in J$ with $\lambda_j > 0$, $\lambda_k < 0$, or (2) there exist $j,k \notin J$ with $\lambda_j > 0$, $\lambda_k < 0$, or (3) after replacing N by $-N$ if necessary so that the signature is positive, there exist $j \in J$ and $k \notin J$ with $\lambda_j > 0$, $\lambda_k > 0$. In any case, the reader may verify that we can choose ϵ small enough so that $a_{IJ} > 0$ when $(I,J) \in \mathcal{P}$ and $a_{IJ} < 0$ when $(I,J) \in \mathcal{N}$. Taking δ sufficiently small, we then obtain

$$Q_b(\phi,\phi) \gtrsim \|\phi\|_{\bar{L}}^2 + \sum_{IJ} |\operatorname{Re}(N\phi_{IJ}, \phi_{IJ})| - (sc)\|\phi\|_L^2 - (\ell c)\|\phi\|^2 .$$

Finally, we use Lemma (5.4.3) together with the fact that $\|\phi\|^2 \leq Q_b(\phi,\phi)$ to conclude

$$\|\phi\|_L^2 + \|\phi\|_{\bar{L}}^2 + \sum_{IJ} |\operatorname{Re}(N\phi_{IJ}, \phi_{IJ})| \lesssim Q_b(\phi,\phi) . \qquad \text{Q.E.D.}$$

We now prove a general theorem from which we shall be able to derive the basic Sobolev estimate for Q_b.

(5.4.7) THEOREM. *Suppose* $A_1, \ldots, A_m$ *are complex vector fields on the (real) manifold* M *such that each* $\bar{A}_j$ *is a linear combination of the* A_j*'s, and suppose that the iterated brackets* $A_i, [A_{i_1}, A_{i_2}], [A_{i_1}, [A_{i_2}, A_{i_3}]], \ldots, [A_{i_1}, [\ldots [A_{i_{p-1}}, A_{i_p}] \ldots]]$ *of order* $\leq p$ *span all vector fields on* M. *Then if* V *is a relatively compact subdomain of* M,

$$\|u\|^2_{2^{1-p}} \lesssim \sum_1^m \|A_j u\|^2 + \|u\|^2$$

uniformly for all smooth functions u *supported in* V.

Proof: Define $F^q_{i_1, \ldots i_q}$ inductively for $1 \leq q \leq p$ by $F^1_i = A_i$, $F^q_{i_1, \ldots i_q} = [A_{i_1}, F^{q-1}_{i_2 \ldots i_q}]$ for $q > 1$. We may assume that V is a coordinate patch so that we are essentially working in a compact subset of Euclidean space. Since the F^q's span all vector fields, we have for any $\epsilon > 0$ and all u supported in V,

$$\|u\|^2_\epsilon \lesssim \sum_{q=1}^p \sum_{i_1 \ldots i_q} \|F^q_{i_1 \ldots i_q} u\|^2_{\epsilon-1} + \|u\|^2 . \tag{5.4.8}$$

At this point we drop the subscripts on the F^q's. If $\Lambda^s = (I - \Delta)^{s/2}$ is the standard elliptic pseudodifferential operator of order s as defined in the Appendix, §1, we have

$$\begin{aligned} \|F^p u\|^2_{\epsilon-1} &\sim (F^p u, \Lambda^{2\epsilon-2} F^p u) = ([A, F^{p-1}]u, \Lambda^{2\epsilon-2} F^p u) \\ &= (AF^{p-1}u, \Lambda^{2\epsilon-2} F^p u) - (F^{p-1} Au, \Lambda^{2\epsilon-2} F^p u) . \end{aligned}$$

Now

$$\begin{aligned} (AF^{p-1}u, \Lambda^{2\epsilon-2} F^p u) &= -(F^{p-1}u, \bar{A}\Lambda^{2\epsilon-2} F^p u) + \mathcal{O}(\|F^{p-1}u\|^2_{2\epsilon-1} + \|u\|^2) \\ &= -(F^{p-1}u, \Lambda^{2\epsilon-2} F^p \bar{A} u) + \mathcal{O}(\|F^{p-1}u\|^2_{2\epsilon-1} + \|u\|^2) ; \end{aligned}$$

here we have used the generalized Schwarz inequality and the fact that $[\bar{A}, \Lambda^{2\epsilon-2}]$ is an operator of order $2\epsilon-2$, cf. Appendix, §1 and §5. Hence

$$|(AF^{p-1}u, \Lambda^{2\epsilon-2}F^p u)| \lesssim \|F^{p-1}u\|_{2\epsilon-1}\|\bar{A}u\| + \mathcal{O}(\|F^{p-1}u\|^2_{2\epsilon-1} + \|u\|^2)$$
$$\lesssim \sum_1^m \|A_j u\|^2 + \|u\|^2 + \|F^{p-1}u\|^2_{2\epsilon-1} \cdot$$

by generalized Schwarz again and the hypothesis on the $\bar{A}$'s. Also,

$$-(F^{p-1}Au, \Lambda^{2\epsilon-2}F^p u) = (Au, \bar{F}^{p-1}\Lambda^{2\epsilon-2}F^p u) + \mathcal{O}(\|Au\|^2_{2\epsilon-1} + \|u\|^2)$$
$$= (Au, \Lambda^{2\epsilon-2}F^p\bar{F}^{p-1}u) + \mathcal{O}(\|Au\|^2_{2\epsilon-1} + \|u\|^2)$$

by the same reasoning, so

$$|(F^{p-1}Au, \Lambda^{2\epsilon-2}F^p u)| \lesssim \sum_1^m \|A_j u\|^2_{\alpha_0} + \|u\|^2 + \|F^{p-1}u\|^2_{2\epsilon-1}$$

where $\alpha_0 = \max(0, 2\epsilon-1)$. Therefore

$$\|F^p u\|^2_{\epsilon-1} \lesssim \sum_1^m \|A_j u\|^2_{\alpha_0} + \|u\|^2 + \|F^{p-1}u\|^2_{2\epsilon-1} \cdot$$

Repeating this argument with p replaced by p−1 and ϵ by 2ϵ, and then continuing inductively, we see that for $0 \leq k \leq p-2$,

$$\|F^{p-k}u\|^2_{2^k\epsilon-1} \lesssim \sum_1^m \|A_j u\|^2_{\alpha_k} + \|u\|^2 + \|F^{p-k-1}u\|^2_{2^{k+1}\epsilon-1}$$

where $\alpha_k = \max(0, 2^{k+1}\epsilon - 1)$. For $k = p-1$ we have simply

$$\|F^1 u\|^2_{2^{p-1}\epsilon-1} = \|Au\|^2_{2^{p-1}\epsilon-1} = (Au, \Lambda^{2^p\epsilon-2}Au)$$
$$\leq \|Au\|\,\|Au\|_{2^p\epsilon-2} \leq \|Au\|_{\alpha_{p-1}}$$

where $\alpha_{p-1} = \max(0, 2^p\epsilon - 2)$. Adding all these inequalities up and substituting in (5.4.8), then, we obtain

$$\|u\|_\epsilon^2 \lesssim \sum_1^m \|A_j u\|_\alpha^2 + \|u\|^2$$

where $\alpha = \max(0, 2^{p-1}\epsilon - 1, 2^p\epsilon - 2)$. Taking $\epsilon = 2^{1-p}$, we are done. Q.E.D.

Remark. With a more intricate proof we could replace $\|u\|_{2^{1-p}}$ by $\|u\|_\epsilon$ for any $\epsilon < 1/p$ in Theorem (5.4.7), cf. Hörmander [19].

We now return to the partially complex manifold X.

(5.4.9) THEOREM. *If* X *satisfies condition* Y(q), *then for all* $\phi \in \mathcal{B}^{p,q}$, $\|\phi\|_{1/2}^2 \lesssim Q_b(\phi,\phi)$.

Proof: In Theorem (5.4.7) we take M to be a subset of X on which $L_1, \ldots, L_{n-1}$ are defined, $\{A_j\} = \{L_i\} \cup \{\overline{L}_i\}$, $p = 2$, and V a set on which the conclusion of Theorem (5.4.6) holds, and we conclude that $\|\phi\|_{1/2}^2 \lesssim \|\phi\|_L^2 + \|\phi\|_{\overline{L}}^2 \lesssim Q_b(\phi,\phi)$ for all $\phi \in \mathcal{B}^{p,q}$ supported in V. The general case now follows by a partition of unity argument, since $Q_b(\zeta\phi, \zeta\phi) \lesssim Q_b(\phi,\phi)$. Q.E.D.

We can now apply the technique of elliptic regularization to deduce existence and regularity theorems for the $\bar\partial_b$ complex analogous to those for the $\bar\partial$-Neumann problem. The estimate of Theorem (5.4.9) takes the place of Theorem (2.4.4); Lemmas (2.4.1, 2, 3, 6) then go through without change with Λ_t^s, Q, and $|||D\phi|||_{s-1}$ replaced by Λ^s, Q_b, and $\|\phi\|_s$, respectively, and we have:

(5.4.10) PROPOSITION. *Suppose* X *satisfies condition* Y(q). *If* $U \subset \overline{U} \subset V \subset X$ *and* $\zeta_1 \in \mathcal{B}^{0,0}$ *is supported in* V, *then for each* $\zeta \in \mathcal{B}^{0,0}$ *supported in* U *and each positive integer* s,

$$\|\zeta\phi\|_{s+1}^2 \lesssim \|\zeta_1(\Box_b + I)\phi\|_s^2 + \|(\Box_b + I)\phi\|^2$$

uniformly for $\phi \in \mathcal{B}^{p,q}$.

(The additional arguments in Theorem (2.4.8) are unnecessary here because there is no "radial" direction.)

We regularize Q_b by adding on an elliptic term as before; the regularity proof for the regularized form proceeds just like the proof of interior regularity in Theorem (2.2.9). Combining this with Proposition (5.4.10), we reproduce the proof in §2.5 to conclude the regularity of Q_b. It then follows that:

(5.4.11) PROPOSITION. *If* X *satisfies condition* Y(q), $\square_b$ *is hypoelliptic, i.e.,* ϕ *is smooth wherever* $\square_b\phi$ *is.*

As in §3.1, we see that the harmonic space $\mathcal{H}_b^{p,q} = \{\phi \in \mathcal{B}^{p,q} : \square_b\phi = 0\}$ is finite-dimensional, and we have the strong orthogonal decomposition

$$H_0^{p,q} = \bar{\partial}_b\vartheta_b \operatorname{Dom}(\square_b) \oplus \vartheta_b\bar{\partial}_b \operatorname{Dom}(\square_b) \oplus \mathcal{H}_b^{p,q}.$$

We define H_b to be orthogonal projection on $\mathcal{H}_b^{p,q}$ and G_b to be the inverse of $\square_b$ on $(\mathcal{H}^{p,q})^{\perp}$ and zero on $\mathcal{H}_b^{p,q}$. (G stands for Green's operator, since the letter N is already being used.) To sum up, we have the analogue of Theorem (3.1.14):

(5.4.12) THEOREM. *Suppose* X *satisfies condition* Y(q). *Then*:

(1) G_b *is a compact operator.*

(2) *For any* $a \in H_0^{p,q}$, $a = \bar{\partial}_b\vartheta_b G_b a + \vartheta_b\bar{\partial}_b G_b a + H_b a$.

(3) $G_b H_b = H_b G_b = 0$; $G_b\square_b = \square_b G_b = I - H_b$ *on* $\operatorname{Dom}(\square_b)$; *and if* N *is also defined on* $H_0^{p,q+1}$ *(resp.* $H_0^{p,q-1}$), $G_b\bar{\partial}_b = \bar{\partial}_b G_b$ *on* $\operatorname{Dom}(\bar{\partial}_b)$ *(resp.* $G_b\vartheta_b = \vartheta_b G_b$ *on* $\operatorname{Dom}(\vartheta_b)$).

(4) $G_b\mathcal{B}^{p,q} \subset \mathcal{B}^{p,q}$, *and for each positive integer* s *the estimate* $\|G_b a\|_{s+1} \lesssim \|a\|_s$ *holds uniformly for* $a \in \mathcal{B}^{p,q}$.

(5.4.13) COROLLARY. *If* $a \in \mathcal{B}^{p,q}$ *then there exists* $\phi \in \mathcal{B}^{p,q-1}$ *such that* $\bar{\partial}_b\phi = a$ *if and only if* $\bar{\partial}_b a = H_b a = 0$. *In that case we may take* $\phi = \vartheta_b G_b a$, *and we have the estimate* $\|\phi\|_s \lesssim \|a\|_s$.

We leave it to the reader to work out the details, as well as the analogues of other propositions in §3.1.

Remarks:

(1) The operator $\Box_b + \overline{\Box}_b$ (where $\overline{\Box}_b\phi = \overline{(\Box_b\overline{\phi})}$) satisfies the "½-estimate"

$$\|\phi\|^2_{1/2} \lesssim ((\Box_b + \overline{\Box}_b + I)\phi, \phi) \tag{5.4.14}$$

in all degrees, provided only that the Levi form does not vanish at any point. Indeed, the argument used to prove Lemma (5.4.5) yields the identity

$$(\overline{\Box}_b\phi,\phi) = \|\phi\|^2_{\overline{L}} - \sum_{IJ}\sum_{k\epsilon J}\lambda_k \operatorname{Re}(N\phi_{IJ},\phi_{IJ}) + R(\phi) + \mathcal{O}(\|\phi\|_L\|\phi\| + \|\phi\|^2)\,.$$

combined with Lemma (5.4.5), this in turn yields

$$\|\phi\|^2_{\overline{L}} + \|\phi\|^2_{\overline{L}} \lesssim ((\Box_b + \overline{\Box}_b + I)\phi, \phi)\,,$$

which implies (5.4.14), by virtue of Theorem (5.4.7). The results of Propositions (5.4.10) and (5.4.11) are therefore valid for $\Box_b + \overline{\Box}_b$.

(2) The case $X = \{z \in C^n : |z| = 1\}$, with the natural partially complex structure induced from C^n, has been investigated in detail by Folland [9], who used the symmetry properties of X to give direct proofs of the "½-estimate" of Theorem (5.4.6) and the global regularity properties of $\overline{\partial}_b$. He also proved an analogue of the Sobolev lemma for $\Box_b + \overline{\Box}_b$. Namely, if we define the norm $|||\ |||_s$ on $\mathcal{B}^{0,0}$ by $|||u|||_s = \|(\Box_b + \overline{\Box}_b + I)^{s/2}u\|$, then $|||\ |||_s$ is stronger than the uniform norm whenever $s > n$. As a consequence, any function u for which $\|(\Box_b + \overline{\Box}_b + I)^{s/2}u\| < \infty$ for some $s > n$ is continuous. (In comparison, the ordinary Sobolev lemma says that if Δ is the Laplace-Beltrami operator, $\|(I - \Delta)^{s/2}(\cdot)\|$ is stronger than the uniform norm if $s > n - 1/2$.)

(3) Let Q be any Hermitian form involving first derivatives defined on sections of a vector bundle over a compact manifold M such that for some $\epsilon > 0$, $\|u\|^2_\epsilon \lesssim Q(u,u)$ for all smooth sections u (and, in case M

has a boundary, $|||Du|||^2_{\epsilon-1} \lesssim Q(u,u)$ near the boundary for all u satisfying suitable boundary conditions). The proof of Theorem (2.4.8) then applies to yield the *a priori estimates* $\|\zeta u\|^2_s \lesssim \|\zeta_1 Fu\|^2_{s-2\epsilon} + \|Fu\|^2$ (or $\|\zeta u\|^2_s \lesssim |||\zeta_1 Fu|||^2_{s-2\epsilon} + \|\zeta_1 Fu\|^2_{s-1} + \|Fu\|^2$) where F is the Friedrichs operator associated to Q. The technique of elliptic regularization then yields the analogue of the Main Theorem for Q; for details, see Kohn and Nirenberg [27]. Operators F associated to such forms Q are called *subelliptic*, and they have recently attracted much attention from Hörmander, Egorov, and others; see, e.g., [19]. (Theorem (5.4.7) provides a useful tool for proving subelliptic estimates.) Thus the $\bar{\partial}_b$ complex has a dual significance: besides its applications to complex analysis, it provides the prototype example of subelliptic behavior.

CHAPTER VI

OTHER METHODS AND RESULTS

1. *The method of weight functions*

After the $\bar{\partial}$-Neumann problem was solved, an alternative approach to the study of the $\bar{\partial}$ complex was developed by L. Hörmander. This method avoids the difficult questions of regularity at the boundary and yields simpler proofs of existence and regularity in the interior. In general, only pseudoconvexity (not necessarily strong) is required; one can even dispense with the assumption that the boundary is smooth. For these reasons, Hörmander's method is in some ways better suited to the needs of the theory of several complex variables than the original version of the $\bar{\partial}$-Neumann problem. However, it provides less information about behavior at the boundary and does not lend itself to the sort of questions discussed in §5.2, 3.

In this section we shall present a brief, proofless sketch of Hörmander's work. To simplify matters, we shall restrict our attention to domains $M \subset \mathbf{C}^n$, for which the metric is flat and questions of cohomology do not arise. For details, proofs, and other results, see Hörmander [16], [18]. Similar techniques have also been used on complete manifolds by Andreotti and Vesentini [4].

We have already had a taste of the weight function technique in the proof of Theorem (3.2.21). The idea is to consider the L^2 norms on forms with respect to the measure $e^{-w}dV$ where dV is Lebesgue measure and w is a real-valued function on M. By suitable choices of w one can then obtain estimates for $\bar{\partial}$ in terms of the w-norms.

To make this precise, let $\|f\|_w = \int_M |f|^2 e^{-w}$ for functions f, and let $H_0^{p,q}(M, w)$ be the space of (p,q)-forms $\phi = \Sigma_{IJ}\phi_{IJ}dz^I \wedge d\bar{z}^J$ with $\|\phi\|_w^2 =$

$\Sigma\|\phi_{IJ}\|_w^2 < \infty$; thus for $w = 0$, $H_0^{p,q}(M,w) = H_0^{p,q}$. We further define $H_s^{p,q}(M, \text{loc})$ to be the space of (p,q)-forms such that $\zeta u \in H_s^{p,q}$ for each $\zeta \in \Lambda_0^{0,0}(M)$. Thus $H_0^{p,q}(M,\text{loc}) = \bigcup_{w \in \mathcal{C}(M)} H_0^{p,q}(M,w)$ ($\mathcal{C}(M)$ = space of continuous real-valued functions on M), and $\Lambda^{p,q}(M) = \bigcap_{s \geq 0} H_s^{p,q}(M,\text{loc})$.

We consider $\bar\partial$ as a closed operator from $H_0^{p,q}(M,w)$ to $H_0^{p,q+1}(M,w)$ and let $\bar\partial^*$ be its Hilbert space adjoint. ($\bar\partial^*$ is given by ϑ + terms of order zero.) The first step is to show that smooth forms are dense in $\text{Dom}\,(\bar\partial) \cap \text{Dom}\,(\bar\partial^*)$ in the graph norm; this is essentially Friedrichs' theorem on weak and strong extensions [10]. It then suffices to obtain estimates for smooth forms. The heart of the matter can be seen in the following proposition, which is easily proved by integration by parts.

(6.1.1) PROPOSITION. *Suppose* bM *is smooth and is given by the equation* $r = 0$. *Then for all smooth forms* $\phi \in \text{Dom}\,(\bar\partial^*)$,

$$\|\bar\partial\phi\|_w^2 + \|\bar\partial^*\phi\|_w^2 = \sum_{kIJ} \left\|\frac{\partial\phi_{IJ}}{\partial\bar z_k}\right\|_w^2 + \sum_{jkIJKL} \epsilon_{jL}^J \epsilon_{kL}^k \int_M \phi_{IJ}\overline{\phi_{IK}} \frac{\partial^2 w}{\partial z_j \partial\bar z_k} e^{-w}$$
$$+ \sum_{jkIJKL} \epsilon_{jL}^J \epsilon_{kL}^k \int_{bM} \phi_{IJ}\overline{\phi_{IK}} \frac{\partial^2 r}{\partial z_j \partial\bar z_k} e^{-w}$$

The first and third terms on the right hand side are familiar to us; it is the second term which allows one to prove estimates by manipulating w. In particular, let w be *strictly plurisubharmonic* in M, i.e., the matrix $\left(\frac{\partial^2 w}{\partial z_i \partial\bar z_j}\right)$ is positive definite. Combining the preceding remarks with some elementary functional analysis, one obtains:

(6.1.2) PROPOSITION. *Suppose* bM *is smooth and M is pseudoconvex, and* w *is a strictly plurisubharmonic function of class* $\mathcal{C}^2$ *on* $\bar M$. *Let* e^{χ} *be the smallest eigenvalue of* $\left(\frac{\partial^2 w}{\partial z_i \partial\bar z_j}\right)$. *Then for every* $\alpha \in H_0^{p,q}(M,w) \cap H_0^{p,q}(M,w+\chi)$ $(q > 0)$ *such that* $\bar\partial\alpha = 0$, *there exists* $\phi \in H_0^{p,q-1}(M,w)$ *such that* $\bar\partial\phi = \alpha$ *and* $\|\phi\|_w^2 \leq \frac{1}{q}\|\alpha\|_{w+\chi}^2$.

We can remove the condition that bM be smooth. In general, we say that a domain M is *pseudoconvex* if there is a plurisubharmonic function σ on M such that $M_s = \{z \in M : \sigma(z) < s\}$ is compact for all real s. (It is not hard to show that this definition coincides with the old one if bM is smooth.) By applying Proposition (6.1.2) to the subdomains M_s and letting $s \to \infty$ one can show:

(6.1.3) PROPOSITION. *Let* M *be a bounded pseudoconvex domain with diameter* D, *and let* w *be plurisubharmonic on* M. *For every* $\alpha \in H_0^{p,q}(M,w)$ $(q>0)$ *with* $\bar\partial\alpha = 0$, *there exists* $\phi \in H_0^{p,q-1}(M,w)$ *such that* $\bar\partial\phi = \alpha$ *and* $\|\phi\|_w^2 \le \frac{eD^2}{q}\|\alpha\|_w^2$.

It is of interest that the estimating constant depends only on the diameter of M.

For any $\alpha \in H_0^{p,q}(M, loc)$ one can find a plurisubharmonic w such that $\alpha \in H_0^{p,q}(M, w)$. Combining this fact with the interior regularity theorems for $\bar\partial$, one obtains:

(6.1.4) PROPOSITION. *If* M *is pseudoconvex, then for any* $\alpha \in H_s^{p,q}(M, loc)$ $(q>0)$ *with* $\bar\partial\alpha = 0$, *there exists* $\phi \in H_{s+1}^{p,q-1}(M, loc)$ *such that* $\bar\partial\phi = \alpha$. *If* $\alpha \in \Lambda^{p,q}(M)$, ϕ *may be chosen in* $\Lambda^{p,q}(M)$.

From Proposition (6.1.4) it is not hard to deduce an alternate proof to the one in §4.2 that a pseudoconvex region is a domain of holomorphy.

A further application of weight functions yields approximation theorems of the Runge type for pseudoconvex domains. If $M \subset \mathbb{C}^n$ is open and $K \subset M$ is compact, K is said to be *pseudoconvex in* M if for every $z \in M - K$ there is a plurisubharmonic function w on M such that $w(z) > 0$ and $w < 0$ on K.

(6.1.5) PROPOSITION. *Suppose* M *is pseudoconvex and* K *is pseudoconvex in* M. *Then any* $\alpha \in H_0^{p,q}(K,0)$ $(q<n)$ *with* $\bar{\partial}\alpha = 0$ *on* K *can be approximated in* $H_0^{p,q}(K,0)$ *by forms* $\phi \in H_0^{p,q}(M, \text{loc})$ *such that* $\bar{\partial}\phi = 0$ *in* M.

Using the fact that the L^2 norm of a holomorphic function on a domain dominates its uniform norm on compact subsets, we obtain the Oka-Weil approximation theorem:

(6.1.6) PROPOSITION. *Suppose* M *is pseudoconvex and* K *is pseudoconvex in* M. *Then any function which is holomorphic in a neighborhood of* K *may be uniformly approximated on* K *by functions which are holomorphic in* M.

All of the preceding results may be extended to domains in Stein manifolds.

Among the other applications of weight function techniques we mention the following: (1) One can obtain existence and approximation theorems for forms on C^n satisfying growth conditions. (2) Extending the theory to manifolds, one obtains finiteness and approximation theorems for the $\bar{\partial}$ cohomology groups of the type discussed in §4.3. (3) Applications to the theory of holomorphic approximation of functions on submanifolds have been developed by Hörmander and Wermer [19a], Nirenberg and Wells [36c], and Harvey and Wells [13c].

2. *Hölder and* L^p *estimates for* $\bar{\partial}$

In this monograph we have only considered estimates for $\bar{\partial}$ in terms of Hilbert space norms; it is natural to inquire what sort of estimates can be obtained for the other common function space norms. Results in this direction have recently been obtained by Grauert and Lieb [13a], Kerzman [20], Lieb [31], Øvrelid [37], Ramirez [38a], and especially Henkin [14].

Here we shall discuss Kerzman's paper [20]; however, we shall be very brief, as the first section of this paper contains an excellent summary of the results as well as references to the work of the other authors mentioned above.

In this section M will denote a relatively compact, strongly pseudoconvex domain in a Hermitian Stein manifold M'. (A complex manifold is Stein if there exists a strictly plurisubharmonic function w on M such that $\{z \in M : w(z) \leq c\}$ is compact for all $c \in \mathbf{R}$.) We may then define L^p norms for forms on M by $\|\phi\|^p_{L^p(M)} = \int_M \langle\phi,\phi\rangle^{p/2}$ if $p < \infty$, and $\|\phi\|_{L^\infty(M)} = \operatorname{ess\,sup}_M \langle\phi,\phi\rangle^{1/2}$. We denote the space of functions (resp. (0,1)-forms) ϕ for which $\|\phi\|_{L^p(M)} < \infty$ by $L^p_0(M)$ (resp. $L^p_1(M)$). We also define Hölder norms for functions u supported in a coordinate patch by $\|u\|_{H^\alpha(M)} = \sup_{z \neq z'} \frac{|u(z) - u(z')|}{|z - z'|^\alpha}$ for $0 < \alpha < 1$, and then define global Hölder norms by means of a fixed partition of unity; the space of functions u for which $\|u\|_{H^\alpha(M)} < \infty$ is denoted by $H^\alpha(M)$. Since $\overline{M}$ is compact, $L^p_k(M) \subset L^q_k(M)$ for $p > q$ $(k = 0,1)$, and $H^\alpha(M) \subset L^\infty_0(M)$ for $0 < \alpha < 1$. The main result is the following:

(6.2.1) PROPOSITION. *For any* $\phi \in L^1_0(M)$ *with* $\bar\partial\phi = 0$, *there exists* $u \in L^1_0(M)$ *such that* $\bar\partial u = \phi$ *and* $\|u\|_{L^1(M)} \lesssim \|\phi\|_{L^1(M)}$. *If* $\phi \in L^p_1(M)$ *then* $u \in L^p_0(M)$ *and* $\|u\|_{L^p(M)} \lesssim \|\phi\|_{L^p(M)}$ $(1 \leq p \leq \infty)$, *and if* $\phi \in \Lambda^{0,1}(M)$ *then* $u \in \Lambda^{0,0}(M)$. *In addition, if* $\phi \in L^\infty_1(M)$ *then* $u \in H^\alpha(M)$ *for* $0 < \alpha < 1/2$, *and* $\|u\|_{H^\alpha(M)} \lesssim \|\phi\|_{L^\infty(M)}$.

The most delicate part of the proof of Proposition (6.2.1) is the construction of local solutions to the equation $\bar\partial u = \phi$ near the boundary which satisfy the L^p and Hölder estimates. Specifically, suppose $x \in bM$ and U is a coordinate neighborhood of x; let $B(x, a)$ be the ball of radius a about x in this coordinate system. Then a local solution u is defined on $M \cap B(x, a)$ for small a by $u(w) = \int_{M \cap B(x,2a)} \Omega(z, w) \wedge \phi(z)$ where $\Omega(z, w)$ is an explicitly constructed $(n,n-1)$-form in z on $M \cap B(x, 2a)$ for each $w \in M \cap B(x, a)$. This local result is combined with the L^2 estimates of Hörmander (cf. §6.1) to yield an extension lemma with bounds, from which the general theorem follows.

Remarks:

(1) If $\phi \in L_1^\infty(M)$ and U is relatively compact in M, one can obtain the estimate $\|u\|_{H^\alpha(U)} \lesssim \|\phi\|_{L^\infty(M)}$ for any $\alpha < 1$, but an example of E. Stein shows that the Hölder estimate for $\alpha > ½$ fails at the boundary of M.

(2) The estimate for $p = 2$ is not a corollary of the $\bar{\partial}$-Neumann problem, since the solution u of Proposition (6.2.1) is not the same as the solution $\vartheta N\phi$ of the $\bar{\partial}$-Neumann problem.

We now give some applications of Proposition (6.2.1).

(6.2.2) PROPOSITION. *If $\phi \in \Lambda^{0,1}(M) \cap L_1^\infty(M)$ and $\bar{\partial}\phi = 0$, there exists $u \in \Lambda^{0,0}(M)$ which is continuous in $\bar{M}$ (even though ϕ may not be) such that $\bar{\partial}u = \phi$.*

(6.2.3) PROPOSITION. *There is a neighborhood $\hat{M}$ of $\bar{M}$ with the following properties:*

(1) *Any continuous function on $\bar{M}$ which is holomorphic in M can be uniformly approximated on $\bar{M}$ by functions holomorphic in $\hat{M}$.*

(2) *If $u \in L_0^p(M)$ $(1 \leq p \leq \infty)$ is holomorphic, there exists a sequence u_n of holomorphic functions on $\hat{M}$ such that $u_n \to u$ uniformly on compact subsets of M, $\|u_n\|_{L^p(M)} \lesssim \|u\|_{L^p(M)}$, and, if $p < \infty$, $u_n \to u$ in $L_0^p(M)$.*

(6.2.4) PROPOSITION. *Suppose $M \subset C^n$. Let A be an arbitrary (possibly empty) subset of bM and let $\alpha : A \to [0, ½)$ be an arbitrary function on A. Let $\mathfrak{F}$ be the sheaf over $\bar{M}$ of germs of functions which are holomorphic in M, continuous in $\bar{M}$, and satisfy a Hölder condition with exponent $\alpha(a)$ at each $a \in A$. Then the first cohomology group $H'(M; \mathfrak{F})$ vanishes.*

3. *Miscellaneous remarks and questions*

We showed in §3.2 that the basic estimate for (p,q)-forms is equivalent to the algebraic condition $Z(q)$. However, the basic estimate is not necessary for the conclusions of the Main Theorem to hold. Indeed, we used the basic estimate only to derive the estimate $|||D\phi|||^2_{-1/2} \lesssim Q(\phi,\phi)$ near bM, but according to the remarks at the end of §5.4, the conclusions of the Main Theorem are valid whenever the subelliptic estimate $|||D\phi|||^2_{\epsilon-1} \lesssim Q(\phi,\phi)$ holds near bM for any $\epsilon > 0$. It would be interesting to know when such weaker subelliptic estimates are valid.

Kohn [25] has recently obtained a result along these lines. Let M be a pseudoconvex (not necessarily strongly pseudoconvex) manifold of dimension 2. Given $p \in bM$, let L be a nonvanishing local section near p of the (one-dimensional) bundle of holomorphic vectors tangent to bM. Let A_m be the space of vector fields near p spanned by $L, \bar{L}$, and their brackets $[L,\bar{L}]$, $[L,[L,\bar{L}]]$, $[\bar{L},[\bar{L},L]], \ldots$, up to order m. p is said to be of *type* m if m is the smallest integer such that $A_m|_p = CT_p(bM)$. If every point of bM is of finite type, we define the *type* of M to be the maximum of the type of p for all $p \in bM$; this number is finite since type is upper semi-continuous and bM is compact. (In particular, M is of type one if and only if it is strongly pseudoconvex.) We then have the following result, the proof of which uses a modification of Theorem (5.4.7):

(6.3.1) PROPOSITION. *If* p *is of type* m, *then the estimate* $|||D\phi|||^2_{2^{-m}-1} \lesssim Q(\phi,\phi)$ *holds for all* $\phi \in \mathfrak{D}^{p,1} \cap \Lambda_0^{p,1}(V \cap \bar{M})$ *where* V *is a sufficiently small neighborhood of* p. *Hence if* M *is of type* m, *the estimate* $|||D\phi|||_{2^{-m}-1} \lesssim Q(\phi,\phi)$ *holds near the boundary for all* $\phi \in \mathfrak{D}^{p,1}$.

Kohn and Nirenberg [27] have also shown that if the form Q is merely *compact* on (p,q)-forms, one can deduce the *a priori* estimate $\|\phi\|^2_s \lesssim \|F\phi\|^2_s$ uniformly for $\phi \in \mathrm{Dom}\,(F) \cap \Lambda^{p,q}(\bar{M})$.

It then follows as in the proof of the Main Theorem that $\phi \in \Lambda^{p,q}(\bar{M})$ whenever $F\phi \in \Lambda^{p,q}(\bar{M})$. (These estimates, as far as we know, are not localizable.) One might therefore ask for conditions under which Q will be compact; however, the usefulness of this question is not clear at present, since experience shows that compactness is usually obtained only as a by-product of subelliptic estimates.

If M is pseudoconvex but the Levi form is identically zero on some region U of bM, it is not hard to show that local regularity fails along U, cf. [25]. However, one can show that if $\alpha \in \Lambda^{p,q}(\bar{M})$ and $\bar{\partial}\alpha = 0$, $H\alpha = 0$, one can find a solution ϕ of $\bar{\partial}\phi = \alpha$ of class $\mathcal{C}^k$ in $\bar{M}$ for any $k < \infty$. We conjecture that ϕ may be chosen in $\Lambda^{p,q}(\bar{M})$.

The theory is also in need of sharper negative results. For example, in §4.2 we constructed an example of a manifold M for which the range of $\bar{\partial}$ is not closed and global regularity for solutions of $\bar{\partial}\phi = \alpha$ fails. In this example the inner boundary of M was strongly pseudoconvex at some points and strongly pseudoconcave at others. It would be of interest to have a general theorem relating such non-uniform behavior of the Levi form to pathologies in the $\bar{\partial}$-Neumann problem.

Or, consider the case where conditions Z(q−1) and Z(q+1) are valid but condition Z(q) fails, i.e., the Levi form has precisely $n-q-1$ positive and q negative eigenvalues at each point. We have shown (Theorem 3.1.19) that in this case we still have good existence and regularity theorems for (p,q)-forms; however, it seems highly likely that $\dim \mathcal{H}^{p,q} = \infty$. This conjecture is at any rate true for functions when M is strongly pseudoconvex, as we remarked in §4.2. Moreover, in this case if $\mathcal{H}^{p,0} \neq 0$ $(p>0)$ then $\dim \mathcal{H}^{p,0} = \infty$, since if $\phi \neq 0 \in \mathcal{H}^{p,0}$ and $f \in \mathcal{H}^{0,0}$ is non-constant, $f^j\phi \in \mathcal{H}^{p,0}$ for each positive integer j, and these forms are linearly independent.

The real-analytic behavior of the $\bar{\partial}$-Neumann problem is as yet not understood. In particular, if M and bM are real-analytic with a real-analytic metric, is the operator F analytic-hypoelliptic? That is, if $F\phi$ is real-analytic in some open $U \subset \bar{M}$, is ϕ also real-analytic in U?

We close this section by posing a topological question about partially complex manifolds. When can the vector field N (which, as we noted, can be globally defined) be normalized so that it generates an action of the circle R/Z on X? If it can, the $\bar{\partial}_b$ complex is *transversally elliptic* in the sense of Atiyah. If the quotient of X by the circle action is a manifold $\tilde{X}$, then $\tilde{X}$ has a natural complex structure, and $\bar{\partial}_b$ is essentially the pullback of $\bar{\partial}$ on $\tilde{X}$. (More precisely, (p,q)-forms on X which transform according to an irreducible representation of the circle can be identified with V-valued (0,q)-forms on $\tilde{X}$ where V is a certain holomorphic vector bundle, and this identification takes $\bar{\partial}_b$ into $\bar{\partial}$.) In case $X = \{z \in C^n : |z| = 1\}$, the circle action exists and is given by scalar multiplication on C^n; the quotient is CP^{n-1}. For a detailed discussion of this case, see Folland [9].

APPENDIX

THE FUNCTIONAL ANALYSIS OF DIFFERENTIAL OPERATORS

1. *Sobolev norms on Euclidean space*

Let $\mathcal{D}$ be the space of $\mathcal{C}^\infty$ functions on $\mathbf{R}^N$ with compact support, $\mathcal{D}(K)$ the space of $\mathcal{C}^\infty$ functions with support in $K \subset \mathbf{R}^N$, and $\mathcal{S}$ the Schwartz space of rapidly decreasing functions on $\mathbf{R}^N$, i.e., the space of all $\mathcal{C}^\infty$ functions u such that $\sup_{\mathbf{R}^N} |x^\alpha D^\beta u| < \infty$ for all multi-indices α and β, where $D^\beta = \left(\frac{1}{i}\frac{\partial}{\partial x}\right)^\beta$. For our present purposes we shall adopt a slightly non-standard metric on $\mathbf{R}^N$, namely $\langle dx_i, dx_j\rangle = \frac{1}{2\pi}\delta_{ij}$, so that the volume element $dx = dx_1 \wedge \ldots \wedge dx_n$ is $(2\pi)^{-N/2}$ times Lebesgue measure; likewise for the dual coordinates $\xi_1, \ldots, \xi_N$. All norms will be taken with respect to this metric. This has the effect of sweeping under the rug all the factors of 2π which customarily plague Fourier analysis.

We define the Fourier transform of a function $u \in \mathcal{S}$ by

$$\mathcal{F}u(\xi) \equiv \hat{u}(\xi) = \int_{\mathbf{R}^N} e^{-i\langle x,\xi\rangle} u(x)dx .$$

We shall assume the following well-known facts about $\mathcal{F}$ (cf. Stein and Weiss [42]):

(1) $\mathcal{F}(D^\beta u)(\xi) = \xi^\beta \hat{u}(\xi)$, $\mathcal{F}(x^\alpha u)(\xi) = -D^\alpha \hat{u}(\xi)$, and if $u_y(x) = u(x+y)$, $\mathcal{F}(u_y)(\xi) = e^{i\langle y,\xi\rangle} u(\xi)$.

(2) $\mathcal{F}$ is a bijection of $\mathcal{S}$ onto itself, and its inverse is given by $\mathcal{F}^{-1}u(x) \equiv \check{u}(x) = \int_{\mathbf{R}^N} e^{i\langle x,\xi\rangle} u(\xi)d\xi$.

(3) (Plancherel theorem) $\mathcal{F}$ extends to a unitary automorphism of $L^2(\mathbf{R}^N)$.

(4) $\mathcal{F}(uv)(\xi) = \int_{\mathbf{R}^N} \hat{u}(\xi-\eta)\hat{v}(\eta)d\eta$.

For any $s \in \mathbf{R}$, we define $\Lambda^s : \mathcal{S} \to \mathcal{S}$ by $\Lambda^s u = \mathcal{F}^{-1}((1+|\xi|^2)^{s/2}\,\hat{u}(\xi))$. (In particular, if s is a positive even integer, by (1) we have $\Lambda^s u = (I-\Delta)^{s/2} u$ where $\Delta = \Sigma_1^N \frac{\partial^2}{\partial x_j^{\,2}}$.) We then define the scalar product $(\,,)_s$ by $(u,v)_s = (\Lambda^s u, \Lambda^s v) = ((1+|\xi|^2)^s \hat{u}, \hat{v})$ and the norm $\|\;\|_s$ by $\|u\|_s = (u,u)_s^{1/2}$. The *Sobolev space* H_s is the completion of $\mathcal{S}$ under the norm $\|\;\|_s$. In particular, $H_0 = L^2(\mathbf{R}^N)$, and we write $\|\;\|_0 = \|\;\|$. Then for all real t, Λ^s extends to a unitary map from H_t onto H_{t-s}. Moreover, H_{-s} may be identified with the dual of H_s under the pairing $\langle u, v\rangle = (\Lambda^s u, \Lambda^{-s} v)$ for $u \in H_s$, $v \in H_{-s}$. The following proposition is trivial from the definitions:

(A.1.1) PROPOSITION (Generalized Schwarz Inequality). *If* $u, v \in H_0$ *and* $u \in H_s$, $s > 0$, *then* $(u,v) \le \|u\|_s \|v\|_{-s}$.

If k is a positive integer it follows from (1) that for any $s \in \mathbf{R}$, $\|u\|_{s+k}^2 \sim \sum_{0 \le |\alpha| \le k} \|D^\alpha u\|_s^2$ for all $u \in \mathcal{S}$. We can then define the *weak (distribution) derivatives* of a function $u \in \mathcal{S}$ as follows: if $\{u_n\} \subset \mathcal{S}$ is a sequence converging to u in H_s, $D^\alpha u$ is the limit in $H_{s-|\alpha|}$ of $D^\alpha u_n$. By the preceding remark, this definition makes sense, for $\{D^\alpha u_n\}$ is Cauchy in $H_{s-|\alpha|}$, and its limit is independent of the choice of the sequence $\{u_n\}$.

If u is a function of class $\mathcal{C}^k$ whose derivatives up to order k are bounded, we define $|u|_k$ to be the supremum over $x \in \mathbf{R}^N$ and $|\alpha| \le k$ of $|D^\alpha u(x)|$. The relation between weak and honest derivatives is then given by the following fundamental theorem.

(A.1.2) PROPOSITION (Sobolev Lemma). $H_s \subset \mathcal{C}^k$ *and* $\|\;\|_s \gtrsim |\;|_k$ *if and only if* $s > k + \frac{N}{2}$. *In this case, if* $u \in H_s$ *then the weak derivatives of* u *up to order* k *are ordinary derivatives.*

Proof: $\|u\|_s \gtrsim |u|_k$ for all $u \in H_s$ if and only if the evaluation functionals E_x^α on $\mathcal{S}$ defined by $E_x^\alpha v = D^\alpha v(x)$ extend to functionals on H_s which are bounded uniformly in x for $|\alpha| \leq k$. But $E_x^\alpha v = \int_{R^N} e^{i<x,\xi>}\xi^\alpha \hat{v}(\xi)d\xi$, so this happens precisely when $e^{i<x,\xi>}\xi^\alpha(1+|\xi|^2)^{-s/2}$ is square-integrable uniformly in x for $|\alpha| \leq k$. Clearly the L^2-norm of this function is independent of x, and by integrating in polar coordinates we see that it is finite if and only if $2|\alpha| - 2s + N - 1 < -1$, i.e., when $s > |\alpha| + \frac{N}{2}$. Finally, if $s > k + \frac{N}{2}$ and $u \in H_s$, choose a sequence $\{u_n\} \subset \mathcal{S}$ converging to u in H_s. The preceding remarks show that for $|\alpha| \leq k$, the weak derivative $D^\alpha u$ is the uniform limit of $\{D^\alpha u_n\}$, so $D^\alpha u$ is continuous and is a derivative of u in the ordinary sense. Q.E.D.

Before proceeding any further, we prove two very useful lemmas.

(A.1.3) LEMMA. $\left(\frac{1+|\xi|^2}{1+|\eta|^2}\right)^s \leq 2^{|s|}(1+|\xi-\eta|^2)^{|s|}$ *for all* $\xi, \eta \in R^N$ *and each real* s.

Proof: Since $|\xi| \leq |\xi-\eta| + |\eta|$, we have $|\xi|^2 \leq 2(|\xi-\eta|^2 + |\eta|^2)$ and hence $1 + |\xi|^2 \leq 2(1+|\xi-\eta|^2)(1+|\eta|^2)$. If $s \geq 0$ the lemma is proved by raising both sides to the s-th power; if $s < 0$ we apply the same argument with ξ and η reversed and s replaced by $-s$. Q.E.D.

(A.1.4) LEMMA. *Suppose* $K(\xi,\eta)$ *is a bounded measurable function on* $R^N \times R^N$ *such that for some* $c > 0$, $\int|K(\xi,\eta)|d\xi \leq c$ *for all* η *and* $\int|K(\xi,\eta)|d\eta \leq c$ *for all* ξ. *(This will be true, for example, if* $|K(\xi,\eta)| \leq |a(\xi-\eta)|$ *where* a *is bounded and* $a(\xi) = \mathcal{O}(|\xi|^{-M})$ *for sufficiently large* M *as* $|\xi| \to \infty$.*) Then the operator* T *defined by* $Tv(\xi) = \int K(\xi,\eta)v(\eta)\,d\eta$ *is bounded on* H_0.

Proof: By the Schwarz inequality,

$$|Tv(\xi)| \leq \int |K(\xi,\eta)|\,|v(\eta)|d\eta \leq \left(\int |K(\xi,\eta)|d\eta\right)^{1/2}\left(\int |K(\xi,\eta)|\,|v(\eta)|^2 d\eta\right)^{1/2},$$

so

$$\|Tv\|^2 = \int |Tv(\xi)|^2 d\xi \le c \iint |K(\xi,\eta)|\, |v(\eta)|^2 d\eta d\xi$$

$$\le c \int |K(\xi,\eta)| d\xi \int |v(\eta)|^2 d\eta = c^2 \|v\|^2 . \qquad \text{Q.E.D.}$$

Since $\| \ \|_s$ is stronger than $\| \ \|_t$ for $s > t$, there is a natural bounded map $i: H_s \to H_t$ which is the identity on $\mathcal{S}$. i is in fact an injection. For, suppose $\{u_n\}$ is a sequence in $\mathcal{S}$ with $u_n \to u$ in H_s, $u_n \to 0$ in H_t. Then for any $v \in \mathcal{S}$,

$$(u, v)_s = \lim (u_n, v)_s = \lim (u_n, \Lambda^{s-t} v)_t = 0 .$$

Since $\mathcal{S}$ is dense in H_s, we conclude that $u = 0$. Therefore we can identify H_s with a dense subspace of H_t whenever $s > t$.

If we restrict our attention to compact sets, we can say more. If $K \subset R^N$ is compact, let $H_{s,0}(K)$ be the completion of $\mathcal{D}(K)$ under $\| \ \|_s$. As before, there is a natural inclusion map $i: H_{s,0}(K) \to H_{t,0}(K)$ for $s > t$.

(A.1.5) LEMMA. *If* $\{u_n\} \subset \mathcal{D}(K)$ *and* $\{\|u_n\|_s\}$ *is bounded, then for any* $t < s$ *there is a subsequence* $\{u_{n_j}\}$ *which converges in* H_t.

Proof: Choose $\phi \in \mathcal{D}$ such that $\phi = 1$ on K. Then $u_n = \phi u_n$, so $\hat{u}_n(\xi) = \int \hat{\phi}(\xi - \eta)\, \hat{u}(\eta)\, d\eta$. By Lemma (A.1.3) and the Schwarz inequality, we see that

$$(1 + |\xi|^2)^{s/2} |\hat{u}_n(\xi)| \lesssim \int |\hat{\phi}(\xi - \eta)|\, (1 + |\xi - \eta|^2)^{|s|/2}\, |\hat{u}(\eta)|\, (1 + |\eta|^2)^{s/2}\, d\eta$$

$$\lesssim \|\phi\|_{|s|}\, \|u_n\|_s \le \text{const.}$$

By the same argument, $(1 + |\xi|^2)^{s/2} |D^\alpha \hat{u}_n(\xi)| \lesssim \|x^\alpha \phi\|_{|s|}\, \|u_n\|_s \le$ const. Thus the $\hat{u}_n$'s and their derivatives are uniformly bounded on compact sets; it follows from the Arzela-Ascoli theorem that there is a subsequence

$\{\hat{u}_{n_j}\}$ which converges uniformly on compact sets. We claim $\{u_{n_j}\}$ converges in H_t for $t < s$. Indeed, for any $R > 0$,

$$\begin{aligned}\|u_{n_j} - u_{n_k}\|_t^2 &= \int (1+|\xi|^2)^t |\hat{u}_{n_j} - \hat{u}_{n_k}|^2(\xi)d\xi \\ &= \int_{|\xi|\le R} (1+|\xi|^2)^t |\hat{u}_{n_j} - \hat{u}_{n_k}|^2(\xi)d\xi \\ &\quad + \int_{|\xi|>R} (1+|\xi|^2)^{t-s}(1+|\xi|^2)^s |\hat{u}_{n_j} - \hat{u}_{n_k}|^2(\xi)d\xi \\ &\lesssim (1+R^2)^{|t|} \sup_{|\xi|\le R} |\hat{u}_{n_j} - \hat{u}_{n_k}|^2(\xi) + (1+R^2)^{t-s} \|u_{n_j} - u_{n_k}\|_s^2 .\end{aligned}$$

Given $\epsilon > 0$, since $t < s$ and $\|u_{n_j} - u_{n_k}\|_s \le \text{const}$, we may choose R so large that $(1+R^2)^{t-s}\|u_{n_j} - u_{n_k}\|_s^2 < \frac{\epsilon}{2}$ for all j, k. But then we may choose j, k large enough so that

$$(1+R^2)^{|t|} \sup_{|\xi|\le R} |\hat{u}_{n_j} - \hat{u}_{n_k}|^2(\xi) < \frac{\epsilon}{2} .$$

Therefore $\{u_{n_j}\}$ is Cauchy in H_t, and the lemma is proved. Q.E.D.

(A.1.6) PROPOSITION (Rellich Lemma). *The inclusion* $i : H_{s,0}(K) \to H_{t,0}(K)$ ($s > t$, K compact) *is a compact map.*

Proof: Given a bounded sequence $\{u_n\} \subset H_{s,0}(K)$, choose sequences $\{u_n^j\} \subset \mathfrak{D}(K)$ with $u_n^j \to u_n$ in $H_{s,0}(K)$ and $\|u_n^j\|_s$ bounded. By Lemma (A.1.5), passing to a subsequence if necessary, $\{u_n^n\}$ has a limit u in $H_{t,0}(K)$ as $n \to \infty$. But then

$$\|u_n - u_m\|_t \le \|u_n - u_n^n\|_t + \|u_n^n - u_m^m\|_t + \|u_m^m - u_m\|_t \to 0 \quad \text{as } m, n \to \infty$$

since $\| \ \|_t \le \| \ \|_s$. Thus $\{u_n\}$ is Cauchy in $H_{t,0}(K)$. Q.E.D.

(A.1.7) PROPOSITION. *If* $s > t$, *then for any* $\epsilon > 0$ *there is a neighborhood* V *of* 0 *such that* $\|u\|_t \le \epsilon \|u\|_s$ *for all* $u \in \mathfrak{D}(V)$.

Proof: First suppose $s > 0$, in which case we may assume $t \geq 0$. If the assertion were false, there would exist $\epsilon > 0$ and a sequence $\{u_j\}$ of functions with supp u_j decreasing to 0 such that $\|u_j\|_t = 1$ and $\|u_j\|_s \leq \frac{1}{\epsilon}$. By the Rellich lemma, a subsequence converges in H_t to a limit u, which is a function since $t \geq 0$. On the one hand, $u(x) = 0$ when $x \neq 0$; on the other, $\|u\|_t = \lim \|u_j\|_t = 1$, and this is absurd.

If $s \leq 0$, let V' be a neighborhood of 0 for which $\|u\|_{-s} \leq \epsilon \|u\|_{-t}$ for all $u \in \mathcal{D}(V')$, and choose $V \subset \overline{V} \subset V'$. Then for all $u \in \mathcal{D}(V)$, by the duality of H_t and H_{-t}, $\|u\|_t = \sup |(u,v)|/\|v\|_{-t}$ where the supremum may be taken over $v \in \mathcal{D}(V')$. Thus

$$\|u\|_t = \sup |(u,v)|/\|v\|_{-t} \leq \epsilon \sup |(u,v)|/\|v\|_{-s} = \epsilon \|u\|_s. \quad \text{Q.E.D.}$$

We now prove the crucial fact that multiplication by a function in $\mathcal{S}$ preserves the Sobolev spaces.

(A.1.8) PROPOSITION. *If* $a \in \mathcal{S}$, *then* $\|au\|_s \lesssim \|u\|_s$ *uniformly for* $u \in \mathcal{S}$. *Hence multiplication by* a *extends to a bounded operator on every* H_s.

Proof: $(1+|\xi|^2)^{s/2}\widehat{au}(\xi) = \int \left(\frac{1+|\xi|^2}{1+|\eta|^2}\right)^{s/2} \hat{a}(\xi-\eta)(1+|\eta|^2)^{s/2}\hat{u}(\eta)\,d\eta$.
Let $K(\xi,\eta) = \left(\frac{1+|\xi|^2}{1+|\eta|^2}\right)^{s/2} \hat{a}(\xi-\eta)$ and $v(\eta) = (1+|\eta|^2)^{s/2}\hat{u}(\eta)$. By Lemma (A.1.3), $|K(\xi,\eta)| \lesssim (1+|\xi-\eta|^2)^{|s|/2}|\hat{a}(\xi-\eta)|$. Since $a \in \mathcal{S}$, the hypothesis of Lemma (A.1.4) is satisfied, and hence

$$\|au\|_s^2 = \int \left| \int K(\xi,\eta)v(\eta)\,d\eta \right|^2 d\xi \lesssim \|v\|^2 = \|u\|_s^2. \quad \text{Q.E.D.}$$

We conclude this section by showing the local invariance of the Sobolev spaces under coordinate transformations. This proof is due to L. Hörmander [15].

(A.1.9) LEMMA. *If* $0 < s < 1$, $\|u\|_s^2 \sim \|u\|^2 + \iint |u(x)-u(y)|^2|x-y|^{-N-2s}dx\,dy$ *for all* $u \in \mathcal{S}$.

Proof: Since $s < 1$ and $|u(x)-u(y)|^2 = \mathcal{O}(|x-y|^2)$ as $x \to y$, the double integral converges absolutely. By the Plancherel theorem,

$$\iint |u(x)-u(y)|^2 |x-y|^{-N-2s} dx\, dy = \iint |u(x+y)-u(y)|^2 |x|^{-N-2s} dx\, dy$$

$$= \iint |e^{i<x,\xi>} - 1|^2 |x|^{-N-2s} |\hat{u}(\xi)|^2 dx\, d\xi$$

since the Fourier transform of $u(x+y)-u(y)$ is $\hat{u}(\xi)(e^{i<x,\xi>}-1)$. By making the substitution $x \mapsto Tx$ where T is an orthogonal transformation, we see that $\int |e^{i<x,\xi>}-1|^2 |x|^{-N-2s} dx$ depends only on $|\xi|$, and by making the substitution $x \mapsto ax$ $(a \in \mathbf{R})$, we see that it is homogeneous of degree $2s$. Hence $\int |e^{i<x,\xi>}-1|^2 |x|^{-N-2s} dx = c_s |\xi|^{2s}$ for some constant $c_s > 0$, and so

$$\iint |u(x)-u(y)|^2 |x-y|^{-N-2s} dx\, dy = c_s \int |\xi|^{2s} |\hat{u}(\xi)|^2 d\xi .$$

Adding this equation to the equation $\int |u(x)|^2 dx = \int |\hat{u}(\xi)|^2 d\xi$, we obtain

$$\int (1+|\xi|^{2s}) |\hat{u}(\xi)|^2 d\xi \sim \|u\|^2 + \iint |u(x)-u(y)|^2 |x-y|^{-N-2s} dx\, dy ,$$

and the lemma follows immediately from the inequalities

$$(1+|\xi|^2)^s \le 1+|\xi|^{2s} \le 2(1+|\xi|^2)^s \quad \text{for } 0<s<1. \qquad \text{Q.E.D.}$$

(A.1.10) PROPOSITION. *Let ψ be a diffeomorphism of an open set $V_1 \subset \mathbf{R}^N$ onto an open set $V_2 \subset \mathbf{R}^N$. Let $K_1 \subset V_1$ be compact and $K_2 = \psi(K_1)$. Then for each s, $\|u\|_s \sim \|u \circ \psi\|_s$ uniformly for $u \in \mathcal{D}(K_2)$.*

Proof: It suffices to show $\|u\|_s \lesssim \|u \circ \psi\|_s$, as the reverse inequality follows from considering ψ^{-1}. Let $J_{\psi^{-1}}$ be the Jacobian determinant of ψ^{-1}, $B_1 = \sup_{K_2} |J_{\psi^{-1}}|$, and $B_2 = \sup_{K_1} \frac{|\psi(x)-\psi(y)|}{|x-y|}$. Set $U = u \circ \psi$,

$x' = \psi(x), y' = \psi(y)$. For $s = 0$, the proposition is true since

$$\int |U(x)|^2\,dx \le B_1 \int |u(x')|^2\,dx'.$$

For $0 < s < 1$ we have

$$\begin{aligned}\iint |U(x)-U(y)|^2|x-y|^{-N-2s}dxdy &= \iint |u(x')-u(y')|^2|x-y|^{-N-2s} \\ &\quad |J_{\psi^{-1}}(x')|\,|J_{\psi^{-1}}(y')|dx'dy' \\ &\le B_2^{N+2s}B_1^2 \iint |u(x')-u(y')|^2 \\ &\quad |x'-y'|^{-N-2s}dx'dy',\end{aligned}$$

so for $0 < s < 1$ the proposition follows from Lemma (A.1.10). If $1 \le s < 2$, we apply this argument to $D^j u$ together with the fact that $\|u\|_s^2 \sim \Sigma \|D^j u\|_{s-1}^2 + \|u\|_{s-1}^2$. Proceeding by induction on the greatest integer in s, we obtain the proposition for all $s \ge 0$.

For $s < 0$ we use the duality of H_s and H_{-s}. Choose functions $\chi_1 \in \mathcal{D}(V_1)$, $\chi_2 \in \mathcal{D}(V_2)$ with $\chi_1 = 1$ on K_1 and $\chi_2 = 1$ on K_2. Then if $v \in \mathcal{S}$,

$$\begin{aligned}\left|\int v\overline{(u\circ\psi)}\right| &= \left|\int (\chi_1 v)\overline{(u\circ\psi)}\right| = \left|\int (\chi_1 v\circ\psi^{-1})\bar{u}|J_\psi|\right| \\ &\le \|(\chi_1 v)\circ\psi^{-1}\|_{-s}\,\|\,|J_\psi|u\|_s \lesssim \|\chi_1 v\|_{-s}\|\chi_2|J_\psi|u\|_s \lesssim \|v\|_{-s}\|u\|_s,\end{aligned}$$

where we have used the generalized Schwarz inequality, the proposition for $-s > 0$, and Proposition (A.1.8). Therefore

$$\|u\circ\psi\|_s = \sup_{v\in\mathcal{S}} \left|\int v(u\circ\psi)\right| / \|v\|_{-s} \lesssim \|u\|_s,$$

and we are done. Q.E.D.

For a more general discussion of Sobolev-type spaces, see Hörmander [15].

2. *Sobolev norms on manifolds*

Suppose X is an N-dimensional Riemannian manifold without boundary. Let $\{U_\gamma\}_{\gamma\epsilon\Gamma}$ be a locally finite covering of X by charts with coordinate mappings $\psi_\gamma : U_\gamma \to \mathbf{R}^N$, and let $\{\zeta_\gamma\}$ be a partition of unity subordinate to $\{U_\gamma\}$. If $K \subset X$ is compact, let $\Lambda_0^k(K)$ be the space of $\mathcal{C}^\infty$ differential forms of degree k supported in K. (If X is itself compact, we take $K = X$ and drop the subscript 0. If X is complex, we have the splitting $\Lambda_0^k(K) = \bigoplus_{p+q=k} \Lambda_0^{p,q}(K)$.) Let $\{\omega_\gamma^I\}_I$ be a pointwise orthonormal basis for the k-forms over U_γ. Then for $s \epsilon \mathbf{R}$ and $\phi \epsilon \Lambda_0^k(K)$ we define $\|\phi\|_s^2 = \Sigma_\gamma \Sigma_I \|(\zeta_\gamma \phi_I^\gamma) \circ \psi_\gamma^{-1}\|_s^2$ where $\phi = \Sigma\, \phi_I^\gamma \omega_\gamma^I$ in U_γ. Note that the sum on the right has only finitely many nonvanishing terms. The completion of $\Lambda_0^k(K)$ with respect to $\|\ \|_s$ is denoted by $H_{s,0}^{\ k}(K)$.

The norm $\|\ \|_s$ is of course highly non-intrinsic. However, Proposition (A.1.8) shows that $\|\ \|_s$ is independent of the choice of local basis and partition of unity up to equivalence. (Given two partitions of unity, the norm defined by each is easily seen to be equivalent to the norm defined by their common refinement.) Also, Proposition (A.1.10) shows that $\|\ \|_s$ is independent of the choice of coordinate mappings, up to equivalence. Therefore, $H_{s,0}^{\ k}(K)$ is well-defined as a topological vector space. Among the equivalent definitions of $\|\ \|_s$ we mention the following.

(1) $\|\ \|_0$ is equivalent to the intrinsic L^2-norm $\|\ \|$ defined by the metric. Indeed, if u is a function supported in K,

$$\|u\|^2 \le \sum_\gamma \|\zeta_\gamma u\|^2 \lesssim \|u\|^2, \quad \|\zeta_\gamma u\|_0^2 = \int_{\mathbf{R}^N} |(\zeta_\gamma u) \circ \psi_\gamma^{-1}|^2 ,$$

and $\|\zeta_\gamma u\|^2 = \int_{\mathbf{R}^N} |(\zeta_\gamma u) \circ \psi_\gamma^{-1}|^2 G$ where $G = \sqrt{\det(g_{ij})}$ is the local expression for the volume density. Since G is smooth and is bounded above and below on $\operatorname{supp} \zeta_\gamma$, the assertion follows immediately.

(2) If s is a positive integer, $\|\phi\|_s^2 \sim \Sigma_\gamma \Sigma_I \Sigma_{|\alpha| \le s} \|D^\alpha[(\zeta_\gamma \phi_I^\gamma) \circ \psi_\gamma^{-1}]\|^2$.

The reader may easily verify that all of the statements of §A.1 still hold for $H_{s,0}^{\ k}(K)$. In particular, the Rellich and Sobolev lemmas are valid.

Now let M be a relatively compact open set in X with smooth boundary bM. We define the norm $\| \ \|_s$ for s a non-negative integer by (2), i.e.

$$\|\phi\|_s^2 = \sum_\gamma \sum_I \sum_{|\alpha|\le s} \|D^\alpha[(\zeta_\gamma \phi_I^\gamma)\circ\psi_\gamma^{-1}]\|^2$$

where the sum is taken over all γ such that $U_\gamma \cap M \neq \emptyset$ and the norms on the right are intrinsic L^2 norms on $\bar{M}$. We denote by $H_s^k(\bar{M})$ the completion of $\Lambda^k(\bar{M})$ with respect to $\| \ \|_s$; as before, $H_s^k(\bar{M})$ is intrinsically defined as a topological vector space. (We could also define $H_s^k(\bar{M})$ for s a negative integer by duality and then for non-integral s by the "quadratic interpolation" functor, but this is unnecessary for our purposes. See Palais [38]; also, see Hörmander [15] for a different approach.)

Proposition (A.1.8) holds for $H_s^k(\bar{M})$ trivially from the definition. To prove the Sobolev and Rellich lemmas we need additional arguments, since their proofs depended strongly on the Fourier transform definition of $\| \ \|_s$. Our plan will be as follows. Let $K \subset X$ be a compact neighborhood of $\bar{M}$; there is a natural continuous restriction map $R: H_{s,0}^k(K) \to H_s^k(\bar{M})$ $(s \in Z^+)$. For each positive integer r we shall construct a right inverse E_r to R which is continuous from $H_s^k(\bar{M})$ to $H_{s,0}^k(K)$ for $0 \le s \le r$. We first prove the local version.

(A.2.1) LEMMA (Lions' Lemma). *Let* $B = \{x \in R^N : |x| < 1\}$, $B^- = \{x \in B : x_N \le 0\}$, $\mathcal{C}_0^r(B^-)$ = *the space of* $\mathcal{C}^r$ *functions on* B^- *with compact support possibly intersecting* $\{x : x_N = 0\}$, $\mathcal{C}_0^r(B)$ = *the space of* $\mathcal{C}^r$ *functions with compact support in* B. *There is a linear map* $E_r : \mathcal{C}_0^r(B^-) \to \mathcal{C}_0^r(B)$ *such that* $E_r u | B^- = u$ *and* $\|E_r u\|_s \lesssim \|u\|_s$ *for* $0 \le s \le r$.

Proof: The matrix $((-j)^m)$ $(1 \le j \le r+1,\ 0 \le m \le r)$ is the Vandermonde matrix of $-1, -2, \ldots, -r-1$ and so is nonsingular; therefore there exist numbers $c_1, \ldots, c_{r+1}$ such that $\Sigma_j c_j (-j)^m = 1$ for $0 \le m \le r$. Define E_r by

$$E_r u(x) = \begin{cases} u(x) & (x_N \leq 0) \\ \sum_i^{r+1} c_j u(x_1, \ldots, x_{N-1}, -jx_N) & (x_N > 0)\,. \end{cases}$$

Then for $x_N > 0$, $D^\alpha E_r u(x) = \Sigma_1^{r+1}(-j)^{\alpha_N} c_j D^\alpha u(x_1, \ldots, x_{N-1}, -jx_N)$. From this it is clear that the left- and right-hand limits of $D^\alpha E_r u$ as $x_N \to 0$ are equal, whence $E_r u \in \mathcal{C}^r$, and that $\|E_r u\|_s \lesssim \|u\|_s$ for $0 \leq s \leq r$. Finally, since $|(x_1, \ldots, x_{N-1}, -jx_N)| \geq |(x_1, \ldots, x_N)|$, supp $E_r u \subset B$. Q.E.D.

(A.2.2) PROPOSITION. *Let* K *be a compact neighborhood of* $\overline{M}$. *For each positive integer* r *there is a bounded linear map* $E_r : H_r^k(\overline{M}) \to H_{r,0}^{\,k}(K)$ *which also extends to a bounded map from* $H_s^k(M)$ *to* $H_{s,0}^{\,k}(K)$ *for* $0 \leq s \leq r$.

Proof: We choose our coordinate covering $\{U_\gamma\}$ so that if $U_\gamma \cap bM \neq \emptyset$, ψ_γ maps $M \cap U_\gamma - \cup_{\beta \neq \gamma} U_\beta$ into $\{x : |x| < ½, x_N < 0\}$, $bM \cap U_\gamma$ into $\{x : x_N = 0\}$, and $U_\gamma - K$ into $\{x : x_N > 1\}$, and we choose the partition of unity $\{\zeta_\gamma\}$ so that supp $(\zeta_\gamma \circ \psi_\gamma^{-1}) \subset \{x : |x| < 1\}$. If $\phi \in \Lambda^k(\overline{M})$, then, we define $E_r^\gamma \phi = \Sigma_I E_r((\zeta_\gamma \phi_I^\gamma) \circ \psi_\gamma^{-1}) \circ \psi_\gamma \omega_I^\gamma$ if $U_\gamma \cap bM \neq \emptyset$ and $E_r^\gamma \phi = \zeta_\gamma \phi$ if $U_\gamma \cap bM = \emptyset$, and set $E_r \phi = \Sigma_\gamma E_r^\gamma \phi$. It is then clear from Lemma (A.2.1) that E_r extends to a bounded map from $H_s^k(\overline{M})$ to $H_{s,0}^{\,k}(K)$ for $0 \leq s \leq r$, as promised. Q.E.D.

(A.2.3) PROPOSITION. *The Sobolev and Rellich lemmas hold on* $\overline{M}$. *More precisely, if* $s > t \geq 0$ *are positive integers, then the inclusion* $i : H_s^k(\overline{M}) \to H_t^k(\overline{M})$ *is compact, and if* $s > \frac{N}{2} + t$, *then every* $\phi \in H_s^k(\overline{M})$ *is* $\mathcal{C}^t$ *on* $\overline{M}$.

Proof: If $\{\phi_n\}$ is a bounded sequence in $H_s^k(\overline{M})$, then $\{E_s \phi_n\}$ is a bounded sequence in $H_{s,0}^{\,k}(K)$. Therefore a subsequence $\{E_s \phi_{n_j}\}$ converges in $H_{t,0}^{\,k}(K)$, whence $\{\phi_{n_j}\}$ converges in $H_t^k(\overline{M})$. If $s > \frac{N}{2} + t$ and $\phi \in H_s^k(\overline{M})$, then $E_s \phi \in H_{s,0}^{\,k}(K)$. Thus the components of $E_s \phi$ are $\mathcal{C}^t$ on K, so their restrictions to $\overline{M}$ are $\mathcal{C}^t$ on $\overline{M}$. Q.E.D.

3. *Tangential Sobolev norms*

In this section we consider smooth functions on $R_-^{N+1} = \{(t,r) \in R^{N+1}: r \leq 0\}$ where $(t_1, \dots, t_N, r)$ are the standard coordinates on R^{N+1} (still with the nonstandard metric, however). We define the *tangential Fourier transform* $\mathcal{F}_t$ by

$$\mathcal{F}_t u(\tau, r) \equiv \tilde{u}(\tau, r) = \int_{R^N} e^{i\langle t, \tau\rangle} u(t, r)\, dt ,$$

the operators Λ_t^s $(s \in R)$ by $\mathcal{F}_t(\Lambda_t^s u)(\tau, r) = (1+|\tau|^2)^{s/2}\tilde{u}(\tau, r)$, and the *tangential Sobolev norms* $|||\ |||_s$ by

$$|||u|||_s^2 = \|\Lambda_t^s u\|^2 = \int_{-\infty}^0 \int_{R^N} (1+|\tau|^2)^s |\tilde{u}(\tau, r)|^2\, d\tau\, dr = \int_{-\infty}^0 \|u(\cdot, r)\|_s^2\, dr .$$

It is clear that the generalized Schwarz inequality and Proposition (A.1.8) still hold for the norms $|||\ |||_s$. The Sobolev and Rellich lemmas are false, since we have no control over derivatives in the r direction. (However, by using Lions' lemma one can show that $|||Du|||_{s-1} \gtrsim \|E_1 u\|_s$ for $0 < s \leq 1$ and u supported in a fixed compact set, which implies that $|||D(\cdot)|||_{s-1}$ is compact with respect to $\|\ \|$ when $s > 0$. This provides an alternative proof of Corollary (3.1.8).)

(A.3.1) PROPOSITION. *If* $s > s'$, *then for any* $\epsilon > 0$ *there exists a neighborhood* V *of* 0 *such that* $|||u|||_{s'} \leq \epsilon |||u|||_s$ *for all* u *supported in* V.

Proof: By Proposition (A.1.7), there is a neighborhood V' of 0 in R^N such that $\|v\|_{s'} \leq \epsilon \|u\|_s$ for all v supported in V'. Let $V = V' \times I$, where I is any interval in $(-\infty, 0]$. Then for all u supported in V,

$$|||u|||_{s'}^2 = \int_{-\infty}^0 \|u(\cdot, r)\|_{s'}^2 dr \leq \epsilon^2 \int_{-\infty}^0 \|u(\cdot, r)\|_s^2\, dr = \epsilon^2 |||u|||_s^2 . \quad \text{Q.E.D.}$$

It is clear that if s is a positive integer, $|||u|||_s^2 \sim \sum_{|\alpha|\le s} \|D_t^\alpha u\|^2$ where the subscript t means that $\alpha_r \equiv \alpha_{N+1} = 0$.

4. *Difference operators*

In this section we prove Lemma (2.2.8) and its analogue at the boundary needed in the proof of Theorem (2.3.4). If u is a function on R^N, we define

$$\Delta_h^j u(x) = \frac{1}{2ih}[u(x+h_j)-u(x-h_j)]$$

where $x \pm h_j = (x_1, \ldots, s_{j-1}, x_j \pm h, x_{j+1}, \ldots, x_N)$. If β is a multi-index and $H = (h_{11}, \ldots, h_{1\beta_1}, \ldots, h_{n1}, \ldots, h_{n\beta_n})$, we set $\Delta_H^\beta = \Pi_{j=1}^n \Pi_{k=1}^{\beta_j} \Delta_{h_{jk}}^j$. If u is a function on R_-^{N+1}, we define $\Delta_H^\beta u$ in the same way with the understanding that only the tangential variables $t_1, \ldots, t_N$ are affected. Δ_H^β is then defined on forms componentwise, but to prove the estimates it suffices to consider functions. Since we are only interested in Δ_H^β as $H \to 0$, we shall assume $|h_{jk}| \le 1$ for all j, k. Let $G(x)\,dx$ (or $G(t,r)\,dt\,dr$) be the volume density defined by the Hermitian metric on the original manifold M. (According to the discussion in §A.2, this will only enter the picture when we compute the adjoint of Δ_H^β.) We recapitulate Lemma (2.2.8):

(A.4.1) PROPOSITION. *Let* K *be a compact set in either* R^N *or* R_-^{N+1}, *and let* u *and* v *be square-integrable functions supported in* K *(in case* $K \subset R_-^{N+1}$, u *and* v *need not vanish on the hyperplane* $r = 0$). *Let* D *be a first-order differential operator and let* $|\beta| = s$. *Then*:

(1) *If* $u \in H_s$, $\|\Delta_H^\beta u\| \lesssim \|u\|_s$ *uniformly as* $H \to 0$.

(2) *If* $u \in H_s$, $\|[D, \Delta_H^\beta]u\| \lesssim \|u\|_s$ *uniformly as* $H \to 0$.

(3) *If* $u \in H_{s-1}$, $(\Delta_H^\beta u, v) = (u, \Delta_H^\beta v) + \mathcal{O}(\|u\|_{s-1}\|v\|)$ *uniformly as* $H \to 0$.

(4) *If* $u \in H_{s-1}$ *and* $v \in H_1$, $|(u, [\Delta_H^\beta, D]v)| \lesssim \|u\|_{s-1}\|v\|_1$ *uniformly as* $H \to 0$.

(5) *If* $u \in H_{s'}$ *and* $\|\Delta_H^\beta u\|_{s'}$ *is bounded as* $H \to 0$ *then* $D^\beta u \in H_{s'}$.

For the proof, we shall need several preliminary results. Until further notice we shall assume $K \subset R^N$.

(A.4.2) LEMMA. $\mathcal{F}(\Delta_h^j u)(\xi) = \dfrac{\sin h\xi_j}{h}\,\hat{u}(\xi)$.

Proof: Immediate from the properties of $\mathcal{F}$. Q.E.D.

(A.4.3) LEMMA. *If* $u \in H_{s'}$, *then for any* $r \leq s'-s$, $\|\Delta_H^\beta u\|_r \leq \|D^\beta u\|_r$.

Proof: Apply Lemma (A.4.2) and the fact that $\left|\dfrac{\sin h\xi_j}{h}\right| \leq |\xi_j|$. Q.E.D.

(A.4.4) LEMMA. *Let* a *be a smooth function of compact support. If* $u \in H_{s'-1}$, *then for any* $r \leq s'-s$, $\|[a, \Delta_H^\beta]u\|_r \lesssim \|u\|_{r+s-1}$ *uniformly as* $H \to 0$.

Proof: It suffices to prove the lemma for $s = 1$, for in general we may commute a through the operators Δ_h^j occurring in Δ_H^β one at a time. The typical term will look like $\Delta_{H'}^{\beta'}[a, \Delta_h^j]\Delta_{H''}^{\beta''}u$ where $\Delta_{H'}^{\beta'}\Delta_h^j\Delta_{H''}^{\beta''} = \Delta_H^\beta$; by Lemma (A.4.3),

$$\|\Delta_{H'}^{\beta'}[a, \Delta_h^j]\Delta_{H''}^{\beta''}u\|_r \leq \|[a, \Delta_h^j]\Delta_{H''}^{\beta''}u\|_{r+|\beta'|} \lesssim \|\Delta_{H''}^{\beta''}u\|_{r+|\beta'|}$$

$$\lesssim \|u\|_{r+|\beta'|+|\beta''|} = \|u\|_{r+s-1}\,.$$

The lemma for $s = 1$, however, is equivalent to the estimate $\|\Lambda^r[a, \Delta_h^j]\Lambda^{-r}v\| \lesssim \|v\|$, where $v = \Lambda^r u$. Now $\mathcal{F}(\Lambda^r[a, \Delta_h^j]\Lambda^{-r}v)(\xi) = \int K(\xi, \eta)\,\hat{v}(\eta)\,d\eta$ where

$$K(\xi, \eta) = \frac{(1+|\xi|^2)^{r/2}}{(1+|\eta|^2)^{r/2}}\left[\frac{\sin h\eta_j}{h} - \frac{\sin h\xi_j}{h}\right]\hat{a}(\xi-\eta)\,.$$

But $\dfrac{\sin h\eta_j}{h} - \dfrac{\sin h\xi_j}{h}$ is bounded for fixed h and equals $\eta_j - \xi_j + \mathcal{O}(h^2)$ as $h \to 0$. Therefore, by Lemma (A.1.3),

$$|K(\xi,\eta)| \lesssim (1+|\xi-\eta|^2)^{|r|/2}|\xi-\eta|\,|\hat{a}(\xi-\eta)| .$$

Since $\hat{a} \in \mathcal{S}$, we apply Lemma (A.1.4) and are done. Q.E.D.

(A.4.5) LEMMA. *If* $u \in H_r$, $r \geq 0$, $(\Delta_h^j u, v) = (u, \Delta_h^j v) + \mathcal{O}(\|u\|_r \|v\|_{-r})$ *uniformly in* h *as* $h \to 0$. *(The scalar products here are taken with respect to the volume density* $G(x)\,dx$.)

Proof: We have

$$\frac{1}{2ih}\int u(x+h_j)\,\bar{v}(x)\,G(x)\,dx = \frac{1}{2ih}\int u(x)\,\bar{v}(x-h_j)\,G(x-h_j)\,dx$$

$$= \frac{-1}{2(-i)h}\int u(x)\,\bar{v}(x-h_j)\,G(x)\,dx$$

$$+ \frac{1}{2ih}\int u(x)\,\bar{v}(x-h_j)[G(x-h_j)-G(x)]\,dx .$$

By the generalized Schwarz inequality, the second term is bounded by $\|\Delta G u\|_r \|v(x-h_j)\|_{-r}$ where $\Delta G = \frac{1}{h}[G(x-h_j)-G(x)]$. But $\|v(x-h_j)\|_{-r} = \|v\|_{-r}$, and $\|\Delta G u\|_r \lesssim \|u\|_r$ by Proposition (A.1.8) since $\Delta G \to \frac{\partial G}{\partial x_j}$ as $h \to 0$ (since u is supported in K, we may truncate ΔG outside a neighborhood of K to obtain a function in $\mathcal{S}$). Likewise,

$$\frac{1}{2ih}\int u(x-h_j)\,\bar{v}(x)\,G(x)dx = \frac{-1}{2(-i)h}\int u(x)\bar{v}(x+h_j)\,G(x)dx + \mathcal{O}(\|u\|_r\|v\|_{-r}) .$$

Adding these results, we are done. Q.E.D.

Proof of Proposition (A.4.1) *when* $K \subset R^N$:

(1) follows directly from Lemma (A.4.3), taking $r = 0$.

(2) We write $D = \Sigma\, a_j D^j$, where we may assume that a_j vanishes outside a neighborhood of K. Since Δ_H^β commutes with D^j, we have $[D, \Delta_H^\beta] = \Sigma\,[a_j, \Delta_H^\beta] D^j$. Applying Lemma (A.4.4) with $r = 0$, $\|[D,\Delta_H^\beta]u\| \lesssim \Sigma\,\|D^j u\|_{s-1} \lesssim \|u\|_s$.

(3) We shall prove a stronger assertion: if $s' \geq 0$ and $u \in H_{s'+s-1}$, then for all $0 \leq r \leq s'$, $(\Delta_H^\beta u, v) = (u, \Delta_H^\beta v) + \mathcal{O}(\|u\|_{r+s-1} \|v\|_{-r})$ uniformly as $H \to 0$. The proof is by induction on s, the initial step of which is provided by Lemma (A.4.5). If $s > 1$, we write $\Delta_H^\beta = \Delta_h^j \Delta_{H'}^{\beta'}$ where $|\beta'| = s-1$. Using the inductive hypothesis with s' replaced by $s'+1$ (since $u \in H_{s'+s-1} = H_{(s'+1)+(s-1)-1}$),

$$(\Delta_H^\beta u, v) = (\Delta_{H'}^{\beta'} u, \Delta_h^j v) + \mathcal{O}(\|\Delta_{H'}^{\beta'} u\|_r \|v\|_{-r})$$

$$= (u, \Delta_{H'}^{\beta'} \Delta_h^j v) + \mathcal{O}(\|\Delta_{H'}^{\beta'} u\|_r \|v\|_{-r} + \|u\|_{r+s-1} \|\Delta_h^j v\|_{-r-1}) .$$

But $\Delta_{H'}^{\beta'} \Delta_h^j = \Delta_h^j \Delta_{H'}^{\beta'} = \Delta_{H'}^\beta$, and by Lemma (A.4.3), $\|\Delta_{H'}^{\beta'} u\|_r \lesssim \|u\|_{r+s-1}$ and $\|\Delta_h^j v\|_{-r-1} \lesssim \|v\|_{-r}$. The proof is therefore complete.

(4) As in the proof of (2), we have $[D, \Delta_H^\beta] = \Sigma [a_j, \Delta_H^\beta] D^j$, so by the generalized Schwarz inequality and Lemma (A.4.4),

$$|(u, [D, \Delta_H^\beta] v)| \lesssim \|u\|_{s-1} \|\sum [a_j, \Delta_H^\beta] D^j v\|_{1-s}$$

$$\lesssim \|u\|_{s-1} \sum \|D^j v\| \lesssim \|u\|_{s-1} \|v\|_1 .$$

(5) From Lemma (A.4.2) and the inequality $\left|\frac{\sin h\xi_j}{h}\right| \leq |\xi_j|$, $\infty > \limsup_{H \to 0} \|\Delta_H^\beta u\|_{s'} = \|D^\beta u\|_{s'}$. Q.E.D.

Proof of Proposition (A.4.1) *when* $K \subset \mathbb{R}^{N+1}_-$: In this case, since Δ_H^β is a tangential operator, Lemmas (A.4.3, 4, 5) remain valid, with the same proofs, if ordinary Sobolev norms are replaced with tangential Sobolev norms. Therefore, by the arguments in the proof for $K \subset \mathbb{R}^N$, we obtain the estimates

$$\|\Delta_H^\beta u\| \lesssim |||u|||_s ,$$

$$\|[D, \Delta_H^\beta] u\| \lesssim \Sigma |||D^j u|||_{s-1}$$

$$(\Delta_H^\beta u, v) = (u, \Delta_H^\beta v) + \mathcal{O}(|||u|||_{s-1} \|v\|),$$

$$|(u, [D, \Delta_H^\beta] v)| \lesssim |||u|||_{s-1} \Sigma \|D^j v\|$$

uniformly as $H \to 0$. (1), (2), (3), and (4) therefore follow from the inequalities $|||u|||_s \lesssim \|u\|_s$ and $\Sigma |||D^j u|||_{s-1} \lesssim \|u\|_s$. Moreover, if $u \in H_{s'}$ and $|\alpha| \le s'$, by Lemma (A.4.2) (with ξ_j replaced by τ_j),

$$\infty > \limsup_{H \to 0} \|\Delta_H^\beta u\|_{s'} \gtrsim \limsup_{H\to 0} \|D^\alpha \Delta_H^\beta u\| = \limsup_{H\to 0} \|\Delta_H^\beta D^\alpha u\|$$
$$= \|D^\beta D^\alpha u\| = \|D^\alpha D^\beta u\| ,$$

which proves (5). Q.E.D.

5. *Operators constructed from* Λ^s *and* Λ_t^s

In this section we prove some estimates for commutators of Λ^s and Λ_t^s with differential operators. These are special cases of general theorems about pseudodifferential operators, for which we refer to Kohn and Nirenberg [26].

An operator $T: \mathcal{S} \to \mathcal{S}$ is said to be of *order* r if for each $s \in R$, $\|Tu\|_s < \|u\|_{s+r}$ for all $u \in \mathcal{S}$. Replacing u by $\Lambda^{-s-r}u$, we see that T is of order r if and only if $\Lambda^s T \Lambda^{-s-r}$ is of order zero for all s, a fact which we shall use repeatedly. Clearly Λ^r and D^α ($|\alpha| \le r$) are of order r; Proposition (A.1.8) says that multiplication by a function is of order zero. (Note that the Fourier transform has no order.)

(A.5.1) **PROPOSITION**. *If* $a \in \mathcal{S}$, *then* $[\Lambda^r, a]$ *is of order* $r-1$, $[\Lambda^r, [\Lambda^r, a]]$ *is of order* $2r-2$, *and* $[[\Lambda^r, a], a]$ *is of order* $r-2$.

Proof: We need to show $\Lambda^s[\Lambda^r, a]\Lambda^{1-r-s}$, $\Lambda^s[\Lambda^r, [\Lambda^r, a]]\Lambda^{2-2r-s}$, and $\Lambda^s[[\Lambda^r, a], a]\Lambda^{2-r-s}$ are of order zero for every s. But

$$\mathcal{F}(\Lambda^s[\Lambda^r, a]\Lambda^{1-r-s}u)(\xi) = \int K_1(\xi, \eta)\hat{u}(\eta)\,d\eta$$

where

$$K_1(\xi, \eta) = \frac{(1+|\xi|^2)^{s/2}}{(1+|\eta|^2)^{(s+r-1)/2}} [(1+|\xi|^2)^{r/2} - (1+|\eta|^2)^{r/2}]\hat{a}(\xi - \eta) .$$

It is easily verified that

$$|(1+|\xi|^2)^{r/2} - (1+|\eta|^2)^{r/2}| \lesssim |\xi-\eta|\,[(1+|\xi|^2)^{(r-1)/2} + (1+|\eta|^2)^{(r-1)/2}]$$

for all ξ, η. We then have, by Lemma (A.1.3),

$$|K_1(\xi,\eta)| \lesssim |\xi-\eta|\left[\left(\frac{1+|\xi|^2}{1+|\eta|^2}\right)^{(s+r-1)/2} + \left(\frac{1+|\xi|^2}{1+|\eta|^2}\right)^{s/2}\right]|\hat{a}(\xi-\eta)|$$

$$\lesssim |\xi-\eta|\,[(1+|\xi-\eta|^2)^{|s+r-1|/2} + (1+|\xi-\eta|^2)^{|s|/2}]\,|\hat{a}(\xi-\eta)|\;.$$

Since $\hat{a} \in \mathcal{S}$, we apply Lemma (A.1.4) and conclude that $\Lambda^s[\Lambda^r, a]\Lambda^{1-r-s}$ is of order zero for all s.

Next, $\mathcal{F}(\Lambda^s[\Lambda^r,[\Lambda^r,a]\Lambda^{2-2r-s}u)(\xi) = \int K_2(\xi-\eta)\hat{u}(\eta)\,d\eta$ where

$$K_2(\xi,\eta) = \frac{(1+|\xi|^2)^{s/2}}{(1+|\eta|^2)^{(s+2r-2)/2}}\,[(1+|\xi|^2)^{r/2} - (1+|\eta|^2)^{r/2}]^2\,\hat{a}(\xi-\eta)\;.$$

The same estimate as above shows that

$$|K_2(\xi,\eta)| \lesssim |\xi-\eta|^2[(1+|\xi-\eta|^2)^{|s+2r-2|/2} + (1+|\xi-\eta|^2)^{|s|/2}]^2\,|\hat{a}(\xi-\eta)|$$

so we apply Lemma (A.1.4) again and conclude that $\Lambda^s[\Lambda^r,[\Lambda^r,a]]\Lambda^{2-2r-s}$ is of order zero for all s.

The last case is a bit more complicated. As a first step, consider the operator T defined by $\mathcal{F}(Tu)(\xi) = \int K_3(\xi,\eta)\hat{u}(\eta)\,d\eta$ where

$$K_3(\xi,\eta) = \sum_1^N \frac{\partial}{\partial\xi_j}(1+|\xi|^2)^{r/2}(\xi_j-\eta_j)\,\hat{a}(\xi-\eta)$$

$$= r\sum_1^N \xi_j(1+|\xi|^2)^{(r-2)/2}(\xi_j-\eta_j)\,\hat{a}(\xi-\eta)\;.$$

T is the "principal part" of $-[\Lambda^r, a]$; we claim that $[\Lambda^r, a] + T$ is of order $r-2$. Indeed, from Taylor's theorem we have

$$(1+|\xi|^2)^{r/2} - (1+|\eta|^2)^{r/2} + \sum_1^N (\xi_j-\eta_j)\frac{\partial}{\partial\xi_j}(1+|\xi|^2)^{r/2} = R(\xi,\eta)$$

where

$$R(\xi,\eta) \lesssim |\xi-\eta|^2 [(1+|\xi|^2)^{(r-2)/2} + (1+|\eta|^2)^{(r-2)/2}] .$$

From our previous representation of $[\Lambda^r, a]$ by Fourier transforms, we then see that $\mathcal{F}(\Lambda^s([\Lambda^r, a]+T)\Lambda^{2-r-s}u)(\xi) = \int K_4(\xi,\eta)\hat{u}(\eta)\,d\eta$ where $K_4(\xi,\eta) = \dfrac{(1+|\xi|^2)^{s/2}}{(1+|\eta|^2)^{(s+r-2)/2}} R(\xi,\eta)\,\hat{a}(\xi-\eta)$. The assertion therefore follows from the same estimates as before.

Now $[[\Lambda^r, a], a] = [[\Lambda^r, a]+T, a] - [T, a]$; since the first term on the right is clearly of order $r-2$, it suffices to show $[T, a]$ is of order $r-2$. Note that $K_3(\xi,\eta) = k(\xi-\eta,\xi)$ where $k(\alpha,\beta) = r\langle\alpha,\beta\rangle(1+|\beta|^2)^{(r-2)/2}\hat{a}(\alpha)$. We then have

$$\mathcal{F}(a\,Tu)(\xi) = \iint \hat{a}(\xi-\tau)\,k(\tau-\eta,\tau)\,\hat{u}(\eta)\,d\tau\,d\eta,$$

and

$$\mathcal{F}(T(au))(\xi) = \iint k(\xi-\tau,\xi)\hat{a}(\tau-\eta)\hat{u}(\eta)d\tau\,d\eta = \iint \hat{a}(\xi-\tau)k(\tau-\eta,\xi)\hat{u}(\eta)\,d\tau\,d\eta$$

where we have replaced τ by $\xi+\eta-\tau$. Therefore

$$\mathcal{F}(\Lambda^s[T, a]\Lambda^{2-r-s}u)(\xi) = \int K_5(\xi,\eta)\,\hat{u}(\eta)\,d\eta$$

where

$$K_5(\xi,\eta) = \frac{(1+|\xi|^2)^{s/2}}{(1+|\eta|^2)^{(s+r-2)/2}} \int \hat{a}(\xi-\tau)\,[k(\tau-\eta,\xi) - k(\tau-\eta,\tau)]\,d\tau .$$

Now

$$|k(\tau-\eta,\xi) - k(\tau-\eta,\tau)| \lesssim |\tau-\eta|\;|\xi-\tau|\;[(1+|\tau|^2)^{(r-2)/2} + (1+|\xi|^2)^{(r-2)/2}]|\hat{a}(\tau-\eta)| ,$$

so $|K_5(\xi,\eta)| \lesssim A_1 + A_2$ where, by Lemma (A.1.3),

$$A_1 = \frac{(1+|\xi|^2)^{s/2}}{(1+|\eta|^2)^{(s+r-2)/2}} \int |\tau-\eta|\,|\xi-\tau|\,|\hat{a}(\tau-\eta)|\,|\hat{a}(\xi-\tau)|\,(1+|\tau|^2)^{(r-2)/2}d\tau$$

$$\lesssim (1+|\xi-\eta|^2)^{|s+r-2|/2} \int |\tau-\eta|\,|\xi-\tau|\,|\hat{a}(\tau-\eta)|\,|\hat{a}(\xi-\tau)|\,(1+|\xi-\tau|^2)^{|r-2|/2}d\tau$$

$$= (1+|\xi-\eta|^2)^{|s+r-2|/2} \int |\tau|\,|\xi-\eta-\tau|\,|\hat{a}(\tau)|\,|\hat{a}(\xi-\eta-\tau)|\,(1+|\xi-\eta-\tau|^2)^{|r-2|/2}d\tau\,,$$

and

$$A_2 = \frac{(1+|\xi|^2)^{s/2}}{(1+|\eta|^2)^{(s+r-2)/2}} \int |\tau-\eta|\,|\xi-\tau|\,|\hat{a}(\tau-\eta)|\,|\hat{a}(\xi-\tau)|\,(1+|\xi|^2)^{(r-2)/2}d\tau$$

$$\lesssim (1+|\xi-\eta|^2)^{|s+r-2|/2} \int |\tau|\,|\xi-\eta-\tau|\,|\hat{a}(\tau)|\,|\hat{a}(\xi-\eta-\tau)|\,d\tau\,.$$

But since $a \in \mathcal{S}$, both of these last integrals are dominated by $\int(1+|\tau|^2)^{-m}(1+|\xi-\eta-\tau|^2)^{-m}d\tau$ for any positive integer m, and by a simple homogeneity argument we see that this expression is dominated by $(1+|\xi-\eta|^2)^{-m+\frac{N}{2}}$. Taking m very large and applying Lemma (A.1.4), we are done. Q.E.D.

(A.5.2) COROLLARY. *If* L *is a differential operator of order* k *with coefficients in* $\mathcal{S}$, *then* $[\Lambda^r, L]$ *is of order* $r+k-1$, $[\Lambda^r,[\Lambda^r,L]]$ *is of order* $2r+k-2$, *and* $[[\Lambda^r,L],L]$ *is of order* $r+2k-2$.

Proof: Let $L = \Sigma a_\alpha D^\alpha$. We have only to observe that Λ^r commutes with D^α, so commutators of Λ^r with L are reduced to commutators of Λ^r with a_α, composed with D^α. Details are left to the reader. Q.E.D.

Finally, we consider operators on R_-^{N+1}. An operator $T: \mathcal{S}(R_-^{N+1}) \to \mathcal{S}(R_-^{N+1})$ (where $\mathcal{S}(R_-^{N+1})$ is the space of restrictions of functions in $\mathcal{S}(R^{N+1})$ to R_-^{N+1}) is said to be of *tangential order* r if $|||Tu|||_s \lesssim |||u|||_{s+r}$ for every $s \in R$. Thus Λ_t^r and D_t^α ($|\alpha|\leq r$) are of tangential order r, and multiplication by $a \in \mathcal{S}(R_-^{N+1})$ is of tangential order zero.

(A.5.3) PROPOSITION. *If* $a \in \mathcal{S}(R_-^{N+1})$, *then* $[\Lambda_t^s, a]$ *is of tangential order* $s-1$, $[\Lambda_t^s, [\Lambda_t^s, a]]$ *is of tangential order* $2s-2$, *and* $[[\Lambda_t^s, a], a]$ *is of tangential order* $s-2$.

Proof: The proof of Proposition (A.5.1) with Λ^r replaced by Λ_t^s yields the necessary estimates for each value of the coordinate r. These estimates all involve $|\tilde{a}(\tau, r)|$, which decreases rapidly as $r \to -\infty$. We may therefore integrate with respect to r and thus prove the proposition. Q.E.D.

We can now prove assertions (1), (2), and (3) of Lemma (2.4.1). Let $A = \zeta_1 \Lambda_t^k \zeta$ where ζ_1 and ζ are smooth functions of compact support with $\zeta_1 = 1$ on supp ζ. Clearly A is of tangential order r. Moreover, if $L = \Sigma a_j D_t^j + b D_r$ is a first-order differential operator (whose coefficients may be assumed to vanish outside a neighborhood of supp ζ_1), Proposition (A.5.3) easily yields the estimates $|||[A,L]u|||_s \lesssim |||Du|||_{s+k-1}$ and $|||[A,[A,L]]u|||_s \lesssim |||Du|||_{s+2k-2}$ since Λ_t^k commutes with D_t^j and D_r; details are left to the reader.

If $G(t, r)$ is the volume density given by the Hermitian metric on M, and $u, v \in \mathcal{S}(R_-^{N+1})$, we have

$$(Au, v) = \int \zeta_1 \Lambda_t^k(\zeta u)\bar{v}G = \int \zeta u \Lambda_t^k(\zeta_1 \bar{v}G)$$

$$= \int u\zeta \Lambda_t^k(\zeta_1 v) G\,dx + \int u\zeta[\Lambda_t^k, G](\zeta_1 \bar{v})\,dx\,,$$

so the formal adjoint of A is $A' = \zeta\Lambda_t^k\zeta_1 + \frac{\zeta}{G}[\Lambda_t^k, G]\zeta_1$. Since G is bounded away from zero on supp ζ, and we may truncate G outside a neighborhood of supp ζ_1 without changing these calculations, it follows from Proposition (A.5.3) that A' is of tangential order k. Moreover,

$$A - A' = \zeta_1\Lambda_t^k\zeta - \zeta\Lambda_t^k\zeta_1 - \frac{\zeta}{G}[\Lambda_t^k, G]\zeta_1$$

$$= \zeta_1[\Lambda_t^k, \zeta] - \zeta[\Lambda_t^k, \zeta_1] - \frac{\zeta}{G}[\Lambda_t^k, G]\zeta_1\,.$$

Thus Proposition (A.5.3) implies that $A - A'$ is of tangential order $k-1$, and that for any first-order differential operator L,

$$|||[A - A', L]u|||_s \lesssim |||Du|||_{s+k-2} \, .$$

REFERENCES

[1] S. Agmon, A. Douglis, and L. Nirenberg, Estimates near the boundary for solutions of elliptic partial differential equations satisfying general boundary conditions, I, *Comm. Pure Appl. Math.* 12 (1959), 623-727; II, ibid. 17 (1964), 35-92.

[2] A. Andreotti, E. E. Levi convexity and H. Lewy problem, *Actes Congrès Intern. Math.* 1970, Gauthier-Villars, Paris.

[3] ——— and C. D. Hill, several articles to appear in *Ann. Scuola Norm. Sup. Pisa.*

[4] ——— and E. Vesentini, Carleman estimates for the Laplace-Beltrami equation on complex manifolds, *Publ. Math. Inst. Hautes Etudes Sci.* 25 (1965), 81-130.

[5] M. E. Ash, The basic estimate of the $\bar{\partial}$-Neumann problem in the non-Kählerian case, *Amer. J. Math.* 86 (1964), 247-254.

[6] L. Bers, *Introduction to several complex variables,* New York University, New York, 1964.

[7] ———, F. John, and M. Schechter, *Partial differential equations,* Interscience, New York, 1964.

[8] P. E. Conner, *The Neumann's problem for differential forms on Riemannian manifolds,* Mem. Amer. Math. Soc. #20, 1956.

[9] G. B. Folland, The tangential Cauchy-Riemann complex on spheres, *Trans. Amer. Math. Soc.,* to appear.

[10] K. Friedrichs, The identity of weak and strong extensions of differential operators, *Trans. Amer. Math. Soc.* 55 (1944), 132-151.

[11] M. P. Gaffney, Hilbert space methods in the theory of harmonic integrals, *Trans. Amer. Math. Soc.* 78 (1955), 426-444.

[12] P. R. Garabedian and D. C. Spencer, Complex boundary value problems, *Trans. Amer. Math. Soc.* 73 (1952), 223-242.

[12a] I. C. Gohberg and M. G. Krein, Systems of integral equations on a half-line with kernels depending on the difference of arguments, *Amer. Math. Soc. Translations* 14 (1960), 217-288.

[13] H. Grauert, On Levi's problem and the imbedding of real analytic manifolds, *Ann. of Math.* 68 (1958), 460-472.

[13a] ______ and I. Lieb, Das Ramirezsche Integral und die Gleichung $\bar{\partial} f = \alpha$ im Bereich der beschränkten Formen, Rice University Studies, to appear.

[13b] S. Greenfield, Cauchy-Riemann equations in several variables, *Ann. Scuola Norm. Sup. Pisa* 22 (1968), 275-314.

[13c] F. R. Harvey and R. O. Wells, Holomorphic approximation and hyperfunction theory on a totally real submanifold of a complex manifold, *Math. Annalen*, to appear.

[14] G. Henkin, (a) Integral representations of holomorphic functions in strongly pseudoconvex domains and certain applications, *Mat. Sbornik* 78 (120): 4 (1969), 611-632 (Russian), English translation in *Math. of the U.S.S.R.*, April 1969, 7(4), 597-616; (b) Integral representations of functions in a strongly pseudoconvex domain and application to the $\bar{\partial}$ problem, *Mat. Sbornik* 82 (124), 2(6), 1970, 300-308 (Russian); (c) Approximation of functions in strongly pseudoconvex domains and a theorem of Z. L. Liebenzon, to appear (Russian); (d) Uniform estimates for solutions of the $\bar{\partial}$ problem in a Weil domain, to appear (Russian).

[15] L. Hörmander, *Linear partial differential operators*, Springer-Verlag, New York, 1963.

[16] ———, L^2 estimates and existence theorems for the $\bar{\partial}$ operator, *Acta Math.* 113 (1965), 89-152.

[17] ———, The Frobenius-Nirenberg theorem, *Arkiv. Mat.* 5 (1965), 425-432.

[18] ———, *An introduction to complex analysis in several variables*, Van Nostrand, Princeton, 1966.

[19] ———, Hypoelliptic second order differential equations, *Acta Math.* 119 (1967), 147-171.

[19a] ——— and J. Wermer, Uniform approximation on compact subsets in $\mathbf{C}^n$, *Math. Scand.* 23 (1968), 5-21.

[20] N. Kerzman, Hölder and L^p estimates for solutions of $\bar{\partial}u = f$ in strongly pseudoconvex domains, *Comm. Pure Appl. Math.* 24 (1971), 301-379.

[21] ———, The Bergman kernel function: differentiability at the boundary, *Math. Annalen.* 195 (1972), 149-158.

[22] J. J. Kohn, Harmonic integrals on strongly pseudoconvex manifolds, I, *Ann. of Math.* 78 (1963), 112-148; II, ibid. 79 (1964), 450-472.

[23] ———, Boundaries of complex manifolds, *Proc. Conference on Complex Manifolds* (Minneapolis), Springer-Verlag, New York, 1965.

[24] ———, Complex vector fields, *Bull. Amer. Math. Soc.* 78 (1972), 1-11.

[25] ———, Boundary behavior of $\bar{\partial}$ on weakly pseudo-convex manifolds of dimension two, *J. Diff. Geom.*, to appear.

[26] ——— and L. Nirenberg, An algebra of pseudo-differential operators, *Comm. Pure Appl. Math.* 18 (1965), 269-305.

[27] ______ and L. Nirenberg, Non-coercive boundary value problems, *Comm. Pure Appl. Math.* 18 (1965), 443-492.

[28] ______ and H. Rossi, On the extension of holomorphic functions from the boundary of a complex manifold, *Ann. of Math.* 81 (1965), 451-472.

[29] ______ and D. C. Spencer, Complex Neumann problems, *Ann. of Math.* 66 (1957), 89-140.

[30] H. Lewy, On the local character of the solutions of an atypical linear differential equation in three variables and a related theorem for regular functions of two complex variables, *Ann. of Math.* 64 (1956), 514-522.

[30a] ______, An example of a smooth linear partial differential equation without solution, *Ann. of Math.* 66 (1957), 155-158.

[31] I. Lieb, (a) Ein Approximationssatz auf streng pseudokonvexen Gebieten, *Math. Annalen* 184 (1969), 55-60; (b) Beschränktheitsaussagen fur den d″-operator, *Nachr. Akad. Wiss. Göttingen, Math. Phys. Kl.*, 1970, 1-7; (c) Die Cauchy-Riemannschen Differentialgleichung auf streng pseudokonvexen Gebieten, to appear.

[32] B. Malgrange, Pseudo-groupes de Lie elliptiques, Séminaire Leray, Collège de France, 1969-70.

[33] C. B. Morrey, Jr., The analytic embedding of abstract real-analytic manifolds, *Ann. of Math.* 68 (1958), 159-201.

[34] ______, *Multiple integrals in the calculus of variations*, Springer-Verlag, New York, 1966.

[35] A. Newlander and L. Nirenberg, Complex analytic coordinates in almost-complex manifolds, *Ann. of Math.* 65 (1957), 391-404.

[36] L. Nirenberg, Remarks on strongly elliptic partial differential equations, *Comm. Pure Appl. Math.* 8 (1955), 648-674.

[36a] ———, A complex Frobenius theorem, *Seminars on Analytic Functions I*, Princeton University Press, 1957.

[36b] R. Nirenberg, On the H. Lewy extension phenomenon, *Trans. Amer. Math. Soc.*, to appear.

[36c] ——— and R. O. Wells, Approximation theorems on differentiable submanifolds of a complex manifold, *Trans. Amer. Math. Soc.* 142 (1969), 15-36.

[37] N. Øvrelid, Integral representation formulas and L^p estimates for the equation $\bar{\partial}u = f$, to appear.

[38] R. S. Palais et al., *Seminar on the Atiyah-Singer index theorem*, Ann. of Math. Studies #57, Princeton University Press, 1965.

[38a] E. Ramirez, Ein Divisionsproblem in der komplexen Analysis mit einer Anwendung auf Randintegraldarstellung, *Math. Annalen* 184 (1970), 172-187.

[39] F. Riesz and B. Sz.-Nagy, *Functional analysis*, Ungar, New York, 1955.

[40] J.-P. Serre, Un théorème de dualité, *Comm. Math. Helv.* 29 (1955), 9-26.

[41] D. C. Spencer, Overdetermined systems of linear partial differential equations, *Bull. Amer. Math. Soc.* 75 (1969), 179-239.

[42] E. M. Stein and G. Weiss, *Introduction to Fourier analysis on Euclidean spaces*, Princeton University Press, 1971.

[43] W. J. Sweeney, The D-Neumann problem, *Acta Math.* 120 (1968), 224-277.

[44] ______, Coerciveness in the Neumann problem, *J. Diff. Geom.* 6 (1972), 375-393.

[45] U. Venugopalkrishna, Fredholm operators on strongly pseudoconvex domains in $\mathbb{C}^n$, to appear.

[46] A. Weil, *Introduction à l'étude des variétés Kähleriennes*, Hermann, Paris, 1958.

[47] H. Weyl, The method of orthogonal projection in potential theory, *Duke Math. J.* 7 (1940), 414-444.

TERMINOLOGICAL INDEX

NOTATIONAL INDEX

Milton Keynes UK
Ingram Content Group UK Ltd.
UKHW010747141123
432541UK00001B/8

9 780691 081205